Seeking New Horizons

Seeking New Horizons

A Perceptual Approach to Geographic Education

HENRY W. CASTNER

McGill-Queen's University Press
Montreal & Kingston • London • Buffalo

© McGill-Queen's University Press 1990
ISBN 0-7735-0728-0

Legal deposit first quarter 1990
Bibliothèque nationale du Québec

Printed in Canada on acid-free paper

This book has been published with the help of a grant
from the Social Science Federation of Canada, using funds
provided by the Social Sciences and Humanities Research
Council of Canada.

Canadian Cataloguing in Publication Data

Castner, Henry W., 1932-
   Seeking new horizons : a perceptual approach to
geographic education
   Includes bibliographical references.
   ISBN 0-7735-0728-0
   1. Geography – Study and teaching. 2. Visual
perception. 3. Cartography – Study and teaching. I.
Title
   G74.c38 1990   910'.7   C89-090350-6

This book was typeset by Typo Data Plus in 10-point
Palatino Roman.

To Claire

Contents

Figures

Acknowledgments

One's intellectual journey is always helped by many others in countless ways. It would be difficult to say whether it was more valuable to have had the opportunity to teach certain topics, to work with others in a research situation, to talk about applications of ideas, or to simply have someone give encouragement. Since all these activities are helpful, there is no rank order implied by the following credits.

All of the students who took my second year introductory courses were, whether they realized it or not, the original testing ground for much that follows. If they only knew, they might be dismayed by how many times my reading an article in a newly published journal led to a lecture, or worse, a new laboratory exercise in which I tried to come to grips with some new idea or finding. Certainly the more direct recipients of these ideas were my graduate students. Fortunately for me, they returned far more than I gave so that I must acknowledge particularly my interactions with Ron Eastman, Roger Wheate, Brian Cromie, Bill Nelson, Gordon Shields, Margaret Sommerville, David Forrest, Keith Connal, Kathy Ford, Roger Barnes, Barry Silic, Melanie Chan, Brian McGregor, Sally Rudd, Ela Rusak-Mazur, and Byron Moldofsky.

On the educational side, I must thank Gary DeLeeuw, Ron Carswell, and Denis Hosford for the opportunity to work on *Thinking About Ontario* and for numerous subsequent occasions where we could exchange, formally and informally, ideas that had developed from that shared experience. In addition, Dick Mansfield, as collaborator in the above project, has been, together with his wife Barbara, a continual source of valuable counsel, feedback, and encouragement.

I also wish to express my appreciation to the Department of Geography, University of Maryland for their generosity and enthusiasm in allowing me to associate with them for the two years during which this book was written. Thanks must also go to Queen's University for its financial assistance and to the Department of Geography for its support and flexibility in making it possible for me to extend my time away from Kingston so as to allow me to continue this project.

There were many hands who contributed to the production of the manuscript and illustrations. Ross Hough provided much useful feedback, suggestions, and skills of execution in producing the many pieces of artwork required for maps and diagrams. At McGill-Queen's University Press, I must thank many helpers. But in particuliar, I wish to acknowledge the thoughtful reactions and many useful suggestions of Colleen Gray and to thank Joan McGilvray.

I am particularly grateful to the Aid to Scholarly Publications Programme, Social Science Federation of Canada, for the grant which has made all of this possible.

Certainly last, but not least, I must thank my wife Claire for the original inspiration for such a "look at geography," for her continual and generous support during the course of this project, and for her reflections on the many ideas that I have been examining.

Seeking New Horizons

Introduction

Perception is the process by which humans acquire knowledge about their environment and about themselves in relation to the environment. Perception is the beginning of knowing, and hence is an essential component of cognition...The information gained through perception is in the world, and it is not the objects, events and things themselves. The information is the structure of what is perceived, and it specifies the world of events, things and objects.

Bikkar Randhawa (1987, 47-8)

...when we really want them to see them, or to see them anew. We look at the whole, and at the parts. We look at the parts in many different orders, trying to see the many ways in which they combine, or fit, or influence each other. We explore the picture or the landscape with our eyes. That is what I would like to ask you to do.

John Holt (1970, 12)

In the past twenty years a great deal of research has appeared concerning the interaction between maps and users of maps. The general hope for this cartographic communication research has been that it could provide some specific results that would be useful in improving the efficiency of the maps we make. Unfortunately, there has been some disappointment with much of that research because it did not provide the specific and detailed design guidance some were seeking. What it did do was to provide more general insights into the communication process and to show map makers how many variables were not under their direct control. In particular, we have come to appreciate how active and crucial a role the map user can play in the communication process.

As a result, we now see map design in the broader contexts of graphic communication, the information needs of map users, and the ways in which users process maps and, by extension, other visual stimuli. What we have not considered is how we can place in those same contexts the ways we educate people about maps and map use. In other words, we have not pursued the idea of people as processors and creators of spatial and environmental information – information that they may wish to analyse and generalize for themselves or to share with others. To do this, these people must have a certain sensitivity to and awareness of the environments around them, of the information within them, and of techniques by which they can represent their ideas about them.

All this suggests an educational agenda that, on one hand, is geographic but, on the other, is perceptual and expressive. This book 1) outlines a rationale for linking visual perception to geographic thought, and 2) describes the basic processes in vision and their logical extensions in concepts that are fundamental to the ways geographers think about the world. Eight perceptual discriminations are shown to lead to eight conceptual goals of geographic thinking; goals that can be examined interactively, starting in the elementary school grades, at increasing levels of complexity and abstraction and with increasingly sophisticated tools throughout the school years in a spiralled approach to the curriculum.

It is my belief that most of the skills that are prominent in elementary geographic curricula relate to the formal conventions of maps, and to the elements and abstract classifications of geographic information. Rarely do we adopt a "user" point of view with children's maps or provide children with simple analytical tools with which to uncover the relationships among elements of geographic information. Nor do we relate our concepts to those in other learning areas. But a spiralled examination of the fundamental goals of geographic thinking, utilizing the innate processes of visual perception, may be a key to developing not only geographic understanding but spatial awareness and graphic literacy as well.

The utilization of perceptual processes in an educational context can be seen most clearly in a technique in music education known as "Orff-Schulwerk." Briefly, this technique can be described as a highly articulated pedagogy that provides experiences for children in the discrimination of the basic perceptual dimensions of sound. This is accomplished by using exploration, imitation, improvization, and creation. By transforming not only auditory but also literary and graphic material into combinations of natural or rhythmic speech, movement, singing, and playing instruments, this technique, in a larger perspective, witnesses to the essential unity of all knowledge. It is a process rather than a product-oriented methodology – one obviously geared to a perceptual approach to listening and music.

The essential relationship of this technique to geographic thinking can be found between improvization in music and "mapping" that emphasizes the character of geographic information and its use in some goal-directed communication rather than simply recording the existence of things in space. I (and thousands of traditional music teachers) have been attracted to this music methodology because of the opportunities it provides children to become active participants in the learning process. Thus music is not taught, it is discovered in its most elemental, natural, and perceptual forms. I suspect the appeal of Orff-Schulwerk also rests on the position in which it places teachers. Instead of being the single source of the right answers, teachers become helpers, facilitators, coordinators, fellow learners, and adjudicators of the admissibility of evidence. As such, Orff-Schulwerk is a much more difficult methodology to implement, but one rich with potential. The rewards range from the obvious development of individuals who are comfortable with active music making to the more subtle ones of learning to cooperate in and contribute to group activity, and to gain confidence in their own abilities as well as to appreciate skills in and contributions by others.

I have limited my explanatory treatment to the one element – improvization – which I believe is the key to adapting such a philosophy into education in the visual realm. I make a case that mapping (and not map making *per se*) is an inherently improvizational activity and that it provides a discovery approach to learning about the world. It is my belief that visual analysis and graphic expression are basic to geographic thinking, and that an improvisational approach to learning about the world rests as naturally in the geographic curriculum as it might in that of art education.

BACKGROUND

This book derives from my teaching, research, and writing over the past two decades about cartographic communication, map design, and map perception. During that time, there have been dramatic changes in the forms and concerns of cartographic practice, particularly as the computer has become more and more a part of the various processes of spatial information management. At the same time, geographers, having to a certain degree rejected maps during the quantitative revolution, have come to view the gathering, analysis, and display of geographic information in far broader terms. As a result, maps (no matter how they are generated) are again regarded as useful alternatives among the various analytical and descriptive tools that geographers use.

Thus, the unity with which cartography and geography might once have been viewed is certainly not clearly seen today. My concern is not to address the question of whether or not the fledgling academic discipline of cartography should go its own way or join forces with some other professional area. Rather I wish to suggest how geographic thinking and mapping (and not map making *per se*) are inextricably intertwined and indeed synonymous. As a result, when speaking of education, the adjectives "cartographic" and "geographic" are nearly interchangeable. My approach here will be a perceptual one, for another by-product of the interest in cartographic communication has been a greater familiarity with the workings of the human visual system, the eye and the brain. Essentially then, I wish to make use of the conceptual unity that exists in looking at maps and in examining the world around us in order to forge an active and exploratory path to geographic education.

The inspirational model that has directed my efforts has been the aforementioned music education method known as Orff-Schulwerk. This method has demonstrated the effectiveness of an experiential approach to understanding music and has suggested by implication the question of whether there is an equivalent approach to understanding in the visual arts. To date, I have not yet discovered the visual equivalent to Orff-Schulwerk and perhaps I will not. But certainly the creative discipline that is involved in cartographic communication may provide the specific ingredient in geographic education that will give it ideal credentials with which to develop such a visual equivalent. Of course, the ultimate condition may be the recognition in elementary education of the essential unity of our interactions with the world around us. In the meantime, it is interesting to discover how some Orff teachers (not to mention the many teachers of dance) are already providing what I believe are some very basic experiences in spatial awareness – an aspect of geographic thinking that is neglected and certainly poorly articulated.

In order to provide a context for my unorthodox approach to geographic education, I have also been looking at other disciplines, some seemingly antithetical as science, art, and music. I have considered these disciplines in light of the nature of the visual system and the results of research in cartographic communication. In doing so, I was struck by the great diversity of ideas, theories, and experiences that appear related to geographic concepts and contexts. I make no apologies for considering these diverse areas, for one's tacit knowledge (Polanyi, 1959,12) is not as formally compartmentalized as the explicit knowledge we work with in academe. In fact, many of my professional activities have led me, at one time or another, to most of these other areas of explicit knowledge. In so doing, my less segregated view of knowledge has been continually reinforced. It is with such a view that one discovers how many fundamental concepts that collectively underlie geographic thinking are shared individually by other disciplines.

My own formal thinking on these matters began with a psychophysical experiment looking at "least-practical differences" in several dimensions of dot area symbols (1964, 1969 with Robinson). Some of the earliest work of my graduate students utilized a visual search methodology. In all these instances, as with similar studies by others, a perfunctory examination was made of what could be termed an "experience factor." In general, it seemed that the more complex the test image utilized, and the more difficult the question posed by the experiment, the more significant was the experience factor in the statistical analysis, and vice versa (Eastman and Castner, 1983). However, despite criticisms (Castner and Cromie, 1982), the question of the nature and significance of this experience factor has never really been addressed.

It seemed, however, that the advantage of experience was something that manifests itself early in a visual encounter (Castner,1979a). But the more time available to a task, or the more time one was willing to spend solving a visual problem, the less was the advantage of experience, whatever it proved to be. Further, if one accepts the convenience of thinking of the visual system as having two complementary processes – foveal and peripheral – then it would seem that the application of experience would be more closely associated with foveal or central vision, and to our conscious thinking about an object of study.

The disturbing question about the visual search methodology was whether the measured differences in subject performance were due to differences in the various test designs or differences in the subjects themselves. In the meantime, work with tachistoscopes (Nelson,1980) and to a certain extent eye-movement recordings (Castner,1973,1978 with Lywood;1984 and 1985 with Eastman) seemed to be a way of neutralizing the experience effect by looking at subjects' initial responses to various map displays in task-specific situations.

However, central to all of these inquiries was the question asked originally by Arthur Robinson (1971): should we be training cartographers to make better maps or map users to be better map readers? One answer to this question has come out of the generally negative reaction to cartographic communication research.[1] People who were looking for specific answers to their own special design problems were put off by the more general findings of much of this research. Indeed, the major realization of this work, collectively viewed, is that map designers have far less control over users' encounters with maps than was promised in the first theoretical discussions of information transmission in the cartographic medium (see Kolacny, 1969 and Ratajski, 1973).

This realization puts a very damp cloth over efforts to provide designers with the kinds of guidance that will assure improved user efficiency from their map products, assuming that is what they really want (Castner,1979b). It does lend greater support to thoughts of looking at the other side of the educational coin – the map users, and the development of ways to make them more flexible and tolerant of new and unexpected design attributes. This is reinforced when one examines current teaching materials which introduce maps, mapping, and geography in the schools (Castner,1984b). The graphic simplicity (but intellectual complexity) of many of these materials suggests that there are many unexplored possibilities for teaching about mapping with which we may be able to improve the ability of map users to contend with the great variety of maps, and the representational and classification problems they raise.

We are also now seeing the results of a great deal of recent research on the brain. In particular, there is increasing evidence that certain motor and movement activities are crucial to the brain's development, and that their neglect can create learning and emotional problems later on. Because of this finding, a case can be made for the retention and even the elaboration of certain music and movement activities in the preschool and elementary school years. The very fact that movements involve experiences in both time and space suggests that there are primitive activities in one's environmental exploration that will prove to have similar value in developing one's spatial awareness. Certainly aspects of geographic thinking utilize both sides of the brain, and evidence of this should be more prominent in our promotion of geography in the school curriculum.

It will eventually come to the attention of the reader that I have not made very many formal references to much of the vast literature on map skills and the· teaching of geography. This should in no way suggest that I have ignored this work of so many teachers and educators. On the contrary, I have been most anxious to make use of this material whenever I could. However, much of what I am trying to establish in this volume is based on a different set of assumptions. As a result, I found that it was often difficult to make use of these materials in positive ways.

To the experts from other disciplines who may come to examine this volume: I hope it is appreciated that I have not attempted to produce comprehensive reviews of particular subject areas. Rather, I have attempted to set, within a context of what I am calling geographic thinking, ideas and theories which I believe can contribute to a more useful and dynamic approach to mapping as an intellectual process and to thinking about and describing our world. I am aware of the possible inadequacies in my extracts of these materials and of the research areas that I may have overlooked. However, I do not wish to be drawn into ongoing disciplinary debates as to the veracity or currency of particular theories. In other words, I am willing, even if the experts are not, to suspend such judgments on a particular idea or concept if doing so provides a better way of organizing our geographic approach to thinking about the world. For me the highest intellectual satisfaction comes not from finding the perfect organizational structure but from finding one which allows me to discover a new way to think about knowledge and the world.

As a result, this volume, by all rights, should be written by a group of scholars, a collective set of specialists in the various subject areas. But until I am able to put down my own ideas about the logical linkages that contribute to some comprehensive whole, I will have to try my best to capture the essence of ideas from these diverse subject areas. I hope that I have succeeded to a satisfactory degree for the purposes of this book. What has been surprising to me are the publication dates of much of the material that I have found useful. Could it be that many new ideas and concepts published two decades ago have simply been ignored? For some there is little evidence of their adoption in the educational literature and published teaching materials in geography. Perhaps they were rejected, but there is little evidence of this either. Thus one is left to surmise that either it simply takes several decades for educators to work out the implications of ideas appearing in the research literature, or, perhaps, it is the practice of classroom teachers to reduce ideas in the literature to simplistic lessons in memorization, or to use maps and atlases only as spatial dictionaries.

In any event, I hope that readers will consider the words of John Holt stated at the outset and provide the necessary leeway in their examination of what may appear to be a disparate and unrelated collection of topics. In this way, perhaps they can discover some of the more exciting aspects of learning about our world.

ORGANIZATION

In brief, this volume is an attempt to explore from a more theoretical and philosophical point of view the implications of cartographic communication research on our approach to geographic education, particularly for the earlier school years. Maps have always been integral to the start of geographic learning. If we are at all alarmed at the present state of geographic knowledge held by our citizens, then we should make sure that this is not related to the way we are using maps in this teaching introduction. In order to pursue this point, I have examined the possibilities of applying in school classrooms a different set of ideas about the way we view the world and our representations of it.

In doing this, I have considered some of the ways in which we experience and describe the world. I have not done this as a critical review or in a definitive way for there is an exhaustive literature in these areas. Rather, I have looked for common ideas and questions that might help in defining our options. This is the essential role of chapters 2 through 7. Each of these chapters examines a particular area of learning or inquiry. Chapter 8 describes and elaborates eight concepts emerging from these first chapters. Viewed collectively, they provide a framework and rationale for a perceptual approach to geography, such as might be taken in designing a textbook.

In somewhat greater detail, the sequence of ideas, starting with chapter 1, goes something like this. The concept of graphicacy has been promoted by some geographers in an effort to add credence to the notion that geographic training with maps enhances the visual literacy of students. Unfortunately, graphicacy has been described in rather narrow terms so that its educational potential appears to be less than forceful. In order to make a stronger case for graphicacy, we must acknowledge its value to other areas of the curriculum. As a "new" educational idea, we must also consider the setting in which it will be introduced – the complex field of education where, in fact, the recipients of this literacy have little say in the order and conduct of things. A perceptual approach to graphicacy would, by definition, speak directly to the needs and skills of this constituency. It is in this context that we can consider the nature of complexity and make the contrast between "graphic complexity" and "intellectual complexity" in graphic images, such as maps. The concept of graphic complexity derives from research in cartographic communication and relates to the ability of viewers to create hierarchies of figures and grounds in graphic images. Intellectual complexity can be thought of as deriving from the processes in mapping whereby information is selected, simplified, classified, and symbolized. Any attempt to make users of maps and other graphic images more discriminating must involve finding ways to help them deal with both the visual and intellectual complexity of the images they examine.

We can, of course, observe infants learning to cope with their environments, even those of some complexity. These are compelling observations and ones on which more productive educational experiences might be built. In other words, the successful perceptual development which we see in young children suggests that our initial educational activities could involve complex graphic images as long as they were also intellectually simple. The very suggestion of introducing children to complex images suggests that we must also be prepared to admit in our examinations of images a variety of answers to the questions we pose. In considering these ideas it is important that we search for the educational activities which best provide students with experiences on which to build mental constructs, associations, and skills for thinking about the world and for managing information about it. Undoubtedly, this kind of search should have an impact upon the ongoing association between academic cartography and geography.

To begin, however, we must have some understanding of the limits within which we can initiate and employ such ideas. One discovers, in chapter 2, that our geographic concepts have not been elaborated in a way that matches the apparent intellectual development of children. Can it be that we have failed to do this, or that the concepts that we traditionally teach cannot easily be broken down for study at the various levels of development? If either is the case, then what skills can we assume children have mastered in their preschool years? A major premise of this volume is that we can rely only upon their innate skills in such areas as visual perception, for the vast majority of children – the sighted ones – have used these skills to derive a stable conception of the world. Thus, in chapter 3, I provide a description of the human visual system with an emphasis on the perceptual aspects of that system which are most important in deriving information from environmental scenes or graphic images of them.

The validity of utilizing these perceptual skills can be seen, I believe, in some methods in music education (chapter 4) where innate experiences in the perception of sounds are utilized to achieve dramatic improvements in both the production and perceptual discrimination of sounds. Such results are gained through the utilization of, among other things, the process of improvization, which involves working with a limited number of perceptual variables so that the child can experience both the variety and the nuance of their interrelationships. These methods raise the question of the possibility of developing an improvizational approach to seeing and to the graphic arts, for it would seem logical that the successful utilization of perceptual skills should not be limited to only one sensory modality. I argue that the concept of mapping is the graphic equivalent of improvization in music in which mapping combines the processes of thinking about some geographic phenomenon, of determining its essential characteristics, of establishing some communication goal for it, of considering the useful forms and modes of its representation, and only then of executing a graphic statement. In other words, mapping focuses more on the character of and relationships among information being represented, and on the uses to which it will be put, than on the simple recording of that information. Implicit in this is the greater attention devoted to the relative positions or contexts of phenomena than to their absolute positions.

Unfortunately, there is little evidence that improvization in this sense has been utilized by educators in the visual arts. Perhaps it is because the communication goals of artistic expression are not as precise as those required by musicians and by cartographers and geographers. I suggest that our communication requirements can bring to graphic education the same discipline found in some music education. The need to establish precisely a visual contrast or gradient in a map or drawing requires an understanding of the syntax of graphic expression. Given this, I will argue in chapter 5 that the solution to these kinds of specific problems may also sharpen the knowing observers' perception of the world, whether they are observing it directly or indirectly through various images.

In chapter 6, I raise the question of improvization in the context of science, for map making is often deemed to be both an art and a science. Could an improvizational approach be taken to science? An affirmative answer seems obvious when the nature of scientific knowledge is examined, for we see that it is essentially a creation of the human mind. Theories are constructed to explain events, and we collect evidence to see if it supports our theory. As long as the weight of the evidence supports the theory, we hold to it; when it does not, we formulate another one. Thus the collection and evaluation of observations in support of arguments is basic to the scientific method. By asking questions with more than one answer, we are establishing a scientific atmosphere in our classrooms. Improvization (or mapping) is the educational vehicle that makes this possible.

In order to focus these ideas on our subject area of geography, it is useful to consider the recent contributions of research in cartographic communication (chapter 7). Central to this research has been a concern for the map user and the specific ways in which he uses spatial information and reacts to its representation (how it is accessed through vision). From this research we have a rich vocabulary describing the ways in which we interact with images and scenes. Taken collectively this vocabulary provides a precise way of developing improvizational techniques that address reading, analysis, and interpretational skills. Thus the emphasis in education can now shift away from teaching about the conventional elements of maps to the more exciting and creative consideration of how we might portray geographic information to support answers to various environmental questions.

It is necessary then to bring all of this to focus in describing what might be included in a perceptual approach to geographic education. Eight concept areas or domains are identified and described in chapter 8. They arise out of the basic discriminations and identifications that we make in vision and lead to the definition of the conceptual goals of geographic thinking – goals that can be sought at all levels of intellectual development.

Thus it is the discrimination of shapes, colors, textures, contexts, and invariant dimensions in scenes or images that allows us to determine such things as the essential characteristics of objects or information, the boundaries between regions, or the homogeneous cores of areas of some unique set of attributes. The formal recognition of such things is basic to geographic activity, and their existence is most often expressed in maps. But the perceptual processes by which they are derived are the same as those we use in witnessing directly the world around us. To use Polanyi's terminology (1959), the explicit knowledge of formal geographic expression can be derived in the same way as the tacit knowledge we hold from having viewed our surroundings. The essential unity of this derivation should, I believe, be a central facet of geographic thinking and education. There, each of these concept areas can be explored in great depth and at many levels of articulation and sophistication. However, since these pedagogic elaborations would make this book prohibitively long, I have not included them. Perhaps one day they will appear as a separate textbook.

CHAPTER ONE

Changing the Educational Relationships Between Cartography and Geography

The healthiest change we could make today, scientifically and socially, as well as theologically, would be to put back into the center the immediate and personal nature of awareness, responsibility, choice and action; to see that all our magnificent science and technology are derived from man, not man from science.   John R. Platt (1968, 19)

When confronted with change, one's competence level drops and adjustment may become a personal threat from which one falls back on old and tried ways. We as teachers and administrators must accept and respond to this very personal factor of threat from change if change is to be accepted and programme to be evolved.   Ron Common (1985)

Traditionally, simple map making and atlas use have been significant, if not the primary, components of geographic education, particularly at the elementary and intermediate school levels. These activities usually revolve around various kinds of "where is" questions (Petchenik, 1979) and thus reinforce an idea that maps serve essentially as repositories of information, that is, as spatial dictionaries. Without any experiences with other kinds of maps, this singular exposure may be in part responsible for the very limited (and certainly simplistic) view of the nature of geography and cartography which is held by much of the general public. In contrast, I suggest that the real intellectual excitement of geography lies in the complexity of the subject, in the challenge of visualizing these complexities (both cartographically and in the mind's eye), and in working out generalizations about them at various levels or scales of expression. All this we might call "geographic thinking." It is a central thesis of this volume that it is in the processes of that thinking, as much as in the products of that thinking, where all the excitement lies.

If this latter view of our disciplines is closer to the truth, then perhaps we should consider other ways for students to begin their cartographic and geographic education in elementary school. I will argue that children beginning their formal studies of the world can, with advantage: (1) work with more complex images, (2) address questions which have more than one answer, and (3) consider attributes of space other than its Euclidean geometry. If this were to be done in repetitive and sequential steps, with carefully chosen tools and feedback, then perhaps we could convey some of the intellectual excitement of geographic thinking without overwhelming students by the complexity of it. In other words, I am suggesting that this excitement derives not so much from knowing the right answer (the capital of Missouri), but rather from making a case for one of a few possible good answers to a question (is the cultural or economic life of that capital influenced more by events in Kansas City or St Louis?). This volume attempts to make a case for an alternate approach to geography and cartography through the definition and elaboration of the idea of geographic thinking.

GRAPHICACY

Among the strongest statements for changing the emphasis of geographic education came from Balchin and Coleman (1965). They proposed the term "graphicacy" to describe the communication of spatial relationships, whether they be with plans, layouts, maps, sketches, or photographs. Subsequently, Balchin (1972) defined graphicacy more precisely as "the communication of spatial information that cannot be conveyed adequately by verbal or numerical means." The choice of the word was also made so that it fit nicely with the other communication terms of literacy (skill in the written language), articulacy or oracy (skill in the spoken language), and numeracy (skill in the use of numbers.)[1] Thus the term gave geographers (and perhaps others) a handle with which to argue for a more central or fundamental place for geography within the school curriculum, as the fourth fundamental language of communication.

Apparently this has generated considerable discussion in the geographic community in the United Kingdom (Boardman, 1983, viii-ix) and may yet lead to more formal recognition of the concept of graphicacy within the British curriculum. It is not clear from Boardman's review of these events just how broadly or narrowly graphicacy is actually being defined and then promoted across the breadth of the curriculum. He does declare (1983,vii) that "graphicacy is the most distinctively geographical form of communication. It is essentially the communication of spatial information through maps and other forms of illustration." On another occasion, he (Boardman, 1976) implies that geography teachers should share with their colleagues in other subjects, especially English and mathematics (and I would add science, history, and art to mention but a few), responsibility for ensuring that graphicacy is developed by all pupils before they leave school. Herein lies a tactical dilemma for geography in elementary education. Any promotion of the entrenchment of more geographic content in the curriculum will meet with very real political difficulties. For this reason the term graphicacy as heretofore defined may be too narrow in its scope and too closely associated with geography to be useful.

On the other hand, teachers with some geographic training may have to be encouraged to take the lead in promoting the teaching of concepts and activities that lay the foundation skills and experiences for all the disciplines that are concerned with the development of environmental awareness and skills in spatial analysis, synthesis, and expression. To do this it may be useful to realize how many other disciplines may also claim a share of the graphicacy idea. In other words, in what ways might english and mathematics (not to mention art and music) see themselves contributing to the development of a broadly defined graphicacy ability in children? Obviously, in English, the verbal description of place, position, direction, and progress through space are ways in which language can compliment graphic expression. Similarly, absolute and relative relationships between objects in space, or of phenomena across space, can be expressed mathematically. But surely, many geographic dimensions of space can also be expressed by the artist. We also owe to psychologists and graphic designers the concepts of planar and hierarchical organization which are so much a part of thematic map design. In terms of experiencing space directly, we may also have to acknowledge the contribution of programs in music and dance education which involve movement in space and the experiences of complex relationships of direction and orientation. Obviously, in dealing with graphicacy, we have a very broad and potentially powerful educational concept, but one that transcends the narrow confines of traditional map skills training.

It should be noted that a number of disciplines may have a stake in the promotion of the concept of graphicacy. What is required then is some agreement on the most fundamental skills and ideas on which such a concept can be built and some sense of a progression of its development. As a child progresses upward, there would be opportunities for increasing specialization or focus, whether it be into such diverse areas as mapping, painting, graphic design, or choreography.

As to what those fundamental skills and ideas might be, one geographer (Boardman, 1985b, 129) considers the four basic map concepts to be direction, location, scale, and symbolism. In contrast, art educators might suggest such fundamentals as form, line, space, distance, color, and texture. It would be difficult, of course, to know just what all these terms really mean, and therefore difficult to compare them. It is my contention, however, that the concepts that we do promote as basic to maps and geography may not be the best ones, for we have not yet broken them down into progressive steps which begin with experiences that are meaningful to young children.

Take, for example, the concept of scale which allows us to measure on a map the distance between two diverse points on the surface of the earth. Of course, to do this correctly requires knowledge of a variety of things such as, for example, the projection system used to make the map, a bit of spherical trigonometry, perhaps some astronomy. But this knowledge comes only after the learning of a long progression of measurement skills that likely started in infancy, when children learn to differentiate between things that can be reached and those that cannot. With mobility, children soon learn to estimate, for example, what can be reached and what cannot, what can be reached first, and so on. All these skills, of course, involve relative measurements of distance. But they form for children the bulk of their experiences in distance measurement when they arrive at the formal part of their education. It is in school that a series of more formal activities are (or should be) undertaken to provide children with an appreciation of what they can already do. In addition, children should learn of the existence of more useful and sophisticated tools for measuring distance. At some point they may arrive at an appreciation for the need to measure distances when taking into account the curved surface of the earth.

Scale is, however, a much broader concept than simply the measurement of distance. It is also a term that conveys to geographers a sense of the level of generalization or the amount of detail in a map. Thus we speak of large- and small-scale maps and can provide examples of both with verbal or mathematical expressions, such as "one inch equals one mile" and "1:10,000,000." On the other hand, we can examine a map and quickly judge whether it is a large-, medium-, or small-scale map, without ever making reference to any formal expression of its scale. We are obviously attending to such collective attributes of the image as the amount of graphic detail and apparent generalization of information, the area of coverage, the kinds of information shown, and so on. While it is sometimes necessary to know the scale of a map precisely expressed in mathematical terms, for example, to make distance measurements), this knowledge is not necessary for a host of other map uses. Perhaps in the long run we would better promote the development of graphicacy if we were to consider the concept of scale through other avenues than strictly through the mathematical relationship between the map and the world. After all, we can only measure distances with reasonable accuracy on the largest scale maps without having to deal with the problems of map projections, the scale factor, and so on. It is shocking that we do not warn young map users of this problem. But I have never seen a school atlas that did. Perhaps we can avoid all this by utilizing initially other ways of teaching complex concepts like scale.

It is not my intention at this point to try and enunciate what progression of concepts and skills might best contribute to the achievement of graphicacy. To date, its place in geography has been very narrowly conceived. There are, however, some ways by which graphicacy could be usefully elaborated. Certainly we could begin by accepting a more flexible definition of what a map might be (or what it can do), rather than restricting ourselves to a large-scale, inventory model. By examining other discipline areas, we may also discover elements, themes, and activities that are common to a broader number of spatial and graphic interests. This would include the consideration of many areas of graphic expression and of the various ways of perceiving the world. In other words, we could study not only the ways we can represent information and ideas, but also the ways we see them, think about them, and imagine them. The very word perception implies an active, not passive, approach to knowledge and learning, and thus intrinsically carries with it great potential for a discovery approach to education.

The inclusion of these elements in our definition of graphicacy would allow the education of our children to consist of helping them to acquire: (1) a level of skill with some tools of graphic analysis, synthesis, and expression, (2) a familiarity with some of the principles of visual perception, and (3) an awareness of their "presence and orientation" in space, both in micro and macrospace. With these abilities, a person should be able to create various graphics that would be useful in thinking about, analysing, describing, and considering a variety of solutions to problems in the world. As well, a person should be able to understand, at least at a rudimentary level, the nature and meaning of most of the types of graphic expression that are commonly found in our contemporary society. Finally, a person should be able to conceptualize space in both Euclidean and non-Euclidean terms, as though it were a rubber blanket on which spatial relationships can change given changing values, attitudes, and costs.

To achieve this kind of mental state among our graphicate students, we must be prepared to entertain changes in our discipline. Unfortunately, it is not always easy to identify what those changes might be nor to assess their possible future impact. Thus any suggestions based on a cause and effect argument are likely to be highly conjectural. It seems more profitable to focus first on the desirable products of change and then on the processes of change. In this way, we can identify where we want to go (our goals) even if our first paths (our processes) toward those goals do not work out as well as we had anticipated. As in steering an automobile, constant adjustments are necessary as rarely do we encounter a perfectly straight road to our destination.

I have little complaint with many of our overall goals in geographic education, but I am concerned as to where many of the paths lead. Many seem to be more convergent than divergent in the opportunities they provide for the student. But before exploring these concerns further, we should acknowledge the magnitude and complexity of the problem we face.

THE SETTING FOR CHANGE

There are three important conditions which will modify the rate and direction of any suggested changes in geographical curriculum: the nature of the educational enterprise itself, the matter of subject complexity, and the changing nature of cartography.

The Complex Field of Education

If we propose some changes in the products and paths of geographic education, there are many groups which may wish to influence the rate and direction of those changes: taxpayers, parents, teachers, boards of education, ministries of education, professional organizations of teachers, geographers, and other professionals, commercial producers of educational materials, and last (and probably least) the students themselves. A proposal for implementing a change will, of course, be viewed in different ways by each of these groups. For discussion purposes, it might be useful to divide them into two groups: the professionals to whom proposals for change are, we must assume, acceptable topics of conversation; and the general public which may have only a limited view of such proposals. While this volume is addressed generally to the former group, I think we should also be concerned with the latter one.

Both geography and cartography have unfortunate images in the minds of the public at large. They are often viewed in the absolute, inventory terms of state capitals and reference maps and not in the relative and dynamic terms of indefinite boundaries, overlapping processes, and the representations of abstract attributes of space itself as expressed in thematic maps. The systematic aspects of geography are relatively invisible compared to the regional ones, particularly the place-name aspect of them. Certainly games such as Trivial Pursuit have helped to reinforce this association of geography and cartography with place names and isolated facts.

One writer even implies that, to a fault, people teaching geography in the lower grades agree in the discipline's innate simplicity (Carswell, 1971). Among the reasons for this, Carswell suggests a gross oversimplification in the minds of teachers as to what is involved in map reading, that the wrong things are being taught about maps, and that in all probability teachers are teaching *about* maps, not *with* maps.

It is in this climate that the public image of geography as a useful discipline and topic of study must be clarified. To encourage more of what the public sees us now doing in schools would appear to be a mistake for, as I perceive it, that image is not seen to be particularly useful. To be more competitive in this age of television and technology, a more dynamic intellectual agenda must be found.

Subject Complexity

A related aspect of this simplistic view is that many things geographic or cartographic appear to be straightforward. So why waste any valuable time on them in school? Certainly many of the maps we supply to children are highly simplistic and overgeneralized in appearance. Indeed, we seem to go to great lengths to avoid providing maps that *look* complex (that resemble the real world in their complexity). What we fail to realize or appreciate is that complexity in maps may be less related to the number of features on a map than to the organization that relates these features (Eastman, 1985,100). Thus many maps intended for children are quite abstract or intellectually complex due to the considerations that were made in leaving out information. These considerations relate primarily to the steps in generalization, such as those necessitated by the choice of scale, and to those involved in the classification and rank ordering of the data classes. As a result, we often provide children with maps that are visually simplistic but intellectually complex, rather than the other way around.

There is some evidence about how children can misinterpret these simplistic map images. In one example, a child reported that there were only three towns in all of New Zealand because only three were shown on the map! Matthews (1986,125) reports on the more general tendency of the youngest school children to personalize their maps, drawing in prominent features that were clearly insignificant to their elders. Thus an early educational problem that must be resolved is the conflict between what adults may think is important and what the children have come to think is important, namely, what they have experienced.

The use of simplistic images stands in sharp contrast to the ways in which infants learn to interpret the world about them and young children learn (as opposed to being taught) their mother's language. We expose infants to complex environments, noisy in both visual and aural dimensions, without having any regard for the "harm" that such exposure might have on them. However, in most cases, with some good feedback, they are able to differentiate smiles from frowns in a matter of months, to navigate their immediate surroundings after a few more months, to speak the language after a year or so, and to read it in four or five. Obviously they were helped in all these accomplishments but they were faced, initially, with some very complex perceptual and cognitive problems.

Can we not also challenge the perceptual powers of children in geographic education? Currently we seem to concentrate on directing "what is this?" or "where is?" questions at the highly generalized and simplified maps found in most children's atlases and school geography books. While these maps are visually quite simple, they are often rather complex intellectually. Their information has been altered, hidden, or left out by some invisible process of generalization and classification. In contrast, we could provide children with relatively complex visual images of the world, such as aerial photographs and very large-scale maps in which the cartographic information has remained relatively unprocessed.[2] With a more complete "true-to-life" image we can help them to deal with the visual complexity problem by making their own classifications and generalizations of the information present.

If we accept these premises, the evidence above suggests that children, before they even show up in our classrooms, are ready and able to confront complex situations. Thus the educational problem, as I see it, is to try to identify where and in what ways we might make use of the more complex images to introduce the intellectual complexity of geographic thinking in some sequential way.

There appear to be two ways to do this. One is by reducing our traditional geographic concepts to their most basic expressions, or to their simplest forms and to sequentially build upon these concepts to reach the more abstract and complex levels. The second is to provide children, at the outset, with tools that will allow them to deal with a complex graphic image or environment. These tools will not be very sophisticated and should depend on the children's innate perceptual ability rather than on their acquired geographic knowledge. I am thinking here of tools such as a simple gray scale which can be used to describe the lightness or darkness of an area on a photographic image. Its use would entail finding the gray-value step which matches that of the area when the two are placed side-by-side. While one would begin with the simple "what-is-this?" or "where-is?" type questions, once the children have mastered the tool the teacher could proceed to explore more difficult problems of classification and generalization while continuing to use the same complex image.

One other advantage of allowing children to work with complex images is that it allows us to entertain questions that have more than one answer. There are several advantages in creating such teaching situations. First, such a method moves us away from the notion that there are always "right" answers. Thus there will be answers that are better or not as good, rather than right or wrong. Facts (measurements in this context) must be presented to support arguments. Second, looking at more than one possible solution to a question leads naturally into an examination of (1) the criteria by which one answer is judged to be more satisfactory than another, and (2) the differences among various measurement procedures. The criteria will relate to the conditions in which the problem arises, thus requiring a greater measure of intellectual logic than absolute geographic knowledge. I believe this should be particularly appealing to teachers who are reluctant to take up new classroom activities without first having something the students do not have – the right answers! Teachers in this context become helpers in making measurements, in arbitrating the admissibility of evidence (measurements), and in learning the logic of arguments. Third, each of the possible answers will inevitably be flawed in some way. The imperfection of any answer thus provides the opportunity for re-addressing the question at a higher or more complex level of measurement or analysis, or at a lower level of generalization. Readiness in this educational situation would thus be defined as the child coming to an understanding of the imperfection of a problem's solution and an appreciation of the need or possibility of another, perhaps more complicated, approach. This seems to me to be a primary educational goal – to generate in the student's mind the realization that there is more to any subject than first meets the eye. Each time we examine a problem or situation, we see it in new ways because of what we learned about it before.

I believe this turns us away from the negative educational value of "wrong" answers, of what Kracht and Boehm (1980) call the "right-wrong syndrome." Children can profitably be challenged to find something that will be useful in supporting an argument or line of reasoning (and not just finding something for its own sake) and in learning to deal with the kinds of errors that arise out of making measurements and of trying out other ideas and solutions to problems.

In more general terms, Mitroff (1984) reaches for the same values when he makes the distinction between puzzles of the Rubik's Cube variety and those involving human beings.[3] Despite all the endearing and fascinating qualities of the former, the latter are radically different in the way they are solved. In addition, societal problems (unlike puzzles) do not have a single, correct solution that is recognized and accepted by all contending parties. Yet despite the significance of problems involving human beings, education in the professions and sciences is preparing students for tackling problems of the Rubik's Cube variety. While utilizing questions with more than one answer does not address all the issues Mitroff is discussing, it is certainly a useful first educational step.

The Changing Nature of Cartography

The production of maps is increasingly becoming an integrated activity starting with the remote acquisition of information in digital form about the earth's surface, the storage and manipulation of that information by computers, and its presentation in hard or ephemeral copy by other computers. In contrast, the traditional and distinctive activities of surveying, geodesy, air photography, remote sensing, photogrammetry, cartography, photography, printing, and so on are collapsing into one grand discipline which we might call "Spatial Information Management." In short, the computer has revolutionized the mapping industry. It may be that printed maps will no longer be an important product line. Instead, an ephemeral image will be produced in which the specific questions of position, context, or connectivity will be addressed.

Unfortunately, many people active in the older disciplines are finding it difficult to adjust their traditional activities to include new operational concepts and skills being created by the use of the new technologies. In addition, those involved with the new technologies are addressing exciting new problems that do not always seem to involve or relate to the older disciplines. As a result, we are in danger of at least dividing, if not splintering, the various disciplines involved at the very time we should be acknowledging their common ground. Thus, understandably, there are many individuals who are concerned with their continuing identity and importance.

Obviously, this is a change which is well upon us and one which is having tremendous impact. While there are undeniable advantages to using computers for many of the complex and repetitive chores in the mapping industry, it is not yet clear how this change should and will affect the "education industry," particularly education in cartography. That the computer is influencing changes in education is undeniable.

But ideally, we should first define some of the changes that might be of educational value, and then see if the computer can be used to produce them, not the other way around.

My concern here, then, is in identifying educational activities that might help students to build the mental constructs and associations which may be useful in thinking about space and spatial relationships, and then in effectively applying the computer in their solving of problems. While spatial information management will continue to be a valuable growth pole for training students in university departments of geography, it does not appear to have provided us with any generically new ways of thinking about the world or solving its problems. Indeed, there is the danger that the technological influence on thinking may be to encourage us to ask computer data banks the simple "where-is?" or "how-much?" types of questions. As a result, we may be fostering the pursuit of right-wrong answers in education at the expense of promoting more diverse thinking skills that might be developed, for example, through the use of a variety of graphic images.

Unfortunately, many of the packaged computer programs developed for elementary education purposes seem little more than baby-sitters that allow children to play some ostensibly geographically or graphically related games. The principal residual benefit from such exposure appears to be in giving children a familiarity with the computer keyboard. Packages for specific map making activities, such as for making projections, are discouragingly simplistic and fail to provide products that could not be as easily obtained in some libraries. It is too bad that so much energy has been invested for such little educational benefit.

In realistic terms, the understanding of the workings and application of any computing system requires exposure to a great deal of math, science, and computer programming. While there is no doubt that many groups want to introduce computers into classrooms at ever lower grade levels, I do not detect any associated enthusiasm among geography teachers to increase the mathematical or scientific components of what they teach in their "geography time." Even at the university level, one can quickly discourage many students by increasing the class time devoted to mathematics, statistics, and computing. Yet it is these very components which provide the analytical skills that are so sought after by employers. For those students who enjoy these subject areas, geography is attractive and rewarding. Our problem is in making the exposure to these areas less onerous.

Perhaps if we first have a clearer idea of what skills are truly valuable in geographic thinking, we will be in a better position to engage the minds of these students. One approach to discovering these skills is through a more thorough definition of the underlying concepts of geographic inquiry. With these clearly defined concepts we may have a better chance of describing a spiralled approach to the development of graphicacy. In so doing, we may provide as well a much-needed tool for vitalizing geographic education.

GEOGRAPHY AND CARTOGRAPHY

In exploring all these ideas, there is the possibility (and some sadly may view it as the danger) that we may wish to rethink the relationships between geography and cartography. This is especially so in the educational area, and in the kinds of things that we do in the name of geography in the classroom. I am using the term "educational" in the mathetic sense (Papert, 1980, 39f), that is, as it pertains to learning. Too much of what I see in schools and universities I would call "training," the passing on of a selection of facts or skills without clearly defining the intellectual reasons for their use or value.

ˉThus I believe we cannot continue to make too exclusive use of a "where is?" approach to geography in the schools. If this is the only activity we can employ to demonstrate the usefulness of maps, then it is time we provided some alternative approaches. This, of course, furnishes an opportunity for further discussion about what geographers in general are trying to discover. Certainly discussions would be valuable within any group of geographers who work together toward some common goals. Perhaps the two groups that would benefit most from an ongoing conversation are those involved in research, and those who work with future teachers of geography.

This is part of the broader question of what the relationship is between questions asked in research and those asked in education, using this latter term in the mathetic sense referred to above. Given the present exponential growth in knowledge, and the cascading of theories to explain various aspects of that knowledge, one wonders if there is ever time to work out the educational implications of a new theory for all levels of education (see, for example, the new approach to vision by David Marr (1982)). Perhaps this is why there seem to be today such weak formal and interactive links between academic and educational departments in universities. Or, one could ask, what is the purpose of theory building in research? Is it simply to provide straw men for further theory building? But perhaps the real value of theories can only be judged *after* we have examined them at various levels of generalization. Certainly I have found it useful to teach at several levels in the university – introductory, upper years, graduate. At each of these levels one is forced to package intellectual concepts and research questions in different ways, at different levels of generalization if you will. Thus in dealing with map projections in an introductory class, I am more concerned with the ways projections distort attributes of the earth's surface. In an upper level class for cartography majors, I am interested in the ways projections are produced. In graduate seminars, I believe it is appropriate to consider ways in which the choice and use of particular projections can be used to convey such concepts as point of view, error, and level of knowledge about places within the resultant map.

In the absence of these discussions between educators and reseachers, perhaps I can offer some points around which they might revolve, and some ideas about how research and education can be related. In any case, I hope that the chapters that follow will help to encourage those discussions, whatever their outcome.

CHAPTER TWO

The Education Connection

By school age, children have come to expect quite arbitrary and, from their point of view, meaningless demands to be placed upon them by adults – the result, most likely, of the fact that adults often fail to recognize the task of conversion necessary to make their questions have some intrinsic significance for the child.

Jerome Bruner (1966, 157)

Good teaching is more a giving of right questions than a giving of right answers.

Joseph Albers (1971, 71)

Instead, we should try to turn out people who love learning so much and learn so well that they will be able to learn whatever needs to be learned.

John Holt (1964, 177)

The educational enterprise in North America is a vast and complex mechanism made up of a variety of groups of different ages, vested interests, and power. For better or worse, much of the collective energy of the enterprise is focused on curriculum – an agenda for what our children will do while they are in school that will contribute to their intellectual growth and mastery of skills. There is also an equally complex body of writing on the theory and practice of children's learning and development. Ideally, a clear picture of the latter would make possible the formulation of a curriculum that leads to the success-ful operation of the overall enterprise. Unfortunately, in the case of geography, the evidence suggests that this formulation has not been made. It is not the goal of this chapter to provide an exhaustive analysis of curriculum or the theory and practice of children's learning and development, but rather to examine particular aspects of research and practice in education that may have contributed to this less than ideal situation. In so doing, there emerge some attractive ways of organizing concepts and activities in cartography and geography that match the stages of children's development of ideas about space and the world.

In attempting to describe the great confusion of aims and objectives of one aspect of a school curriculum, that of science, The Science Council of Canada (Science 36, 26) noted four levels at which such a description might take place. At the ministry of education level, we can talk about the *intended* curriculum. At the school board or school level, we can consider the *planned* curriculum. In the classroom, the *taught* curriculum is of concern. Finally, there is the curriculum that was actually *learned* by the students themselves. But it may be simpler and more manageable to consider the distinction between "(1) what children *can* learn, (2) what children *should* learn, and (3) what children *do* learn" (Petchenik, 1984, 14) within the North American educational context.

The question of *can* relates to what children are capable of doing given some level of perceptual-cognitive development. The work of Jean Piaget, in particular, has provided a formal description of the nature of knowledge that children exhibit at each stage of development; he was considerably less interested in the processes that make growth possible (Bruner, 1966, 7). Unfortunately, in geographic education, we seem only to be concerned with aspects of our subject within Piaget's highest developmental stage, the formal operational stage. However, the two intermediate stages – the pre-operational and concrete operational – provide obvious opportunities for activities which have never been described or implemented. In fact, we seem to have used Piaget's work to focus on what children in particular grades cannot do with some kinds of maps, not on what they could do with other kinds. In contrast, more recent studies have demonstrated that children *are* capable of dealing with maps and map-like objects at very early ages. While we are not yet clear about what skills and concepts children can work with, we must not allow a narrow view of what maps can be to restrict the exploration of a more broadly defined concept of graphicacy.

The question of what children *do* learn involves an examination of the activities undertaken in and around the school. This would in itself be an exhausting if not impossible task. However, some evidence implying what *is* taught will be examined. It is suggested that in the reality of the classroom, a different set of pressures operate which lead to two problems. The first is that geographic skills and concepts may too easily be reduced to activities involving rote memorization. The second is that we have not really identified the true underlying skills and concepts which lead to those we currently do teach.

Finally, we can consider what children *should* do in geography lessons. There are a large number of possibilities of what *might* be taught. If we agree with Petchenik (1984, 14), as I do, that this is basically a question of values and not one appropriate to research, then we must provide some logical basis for compiling such a list. A number of general and specific proposals for what *should* be taught will be made in this and subsequent chapters.

WHAT CHILDREN CAN LEARN

The question of *can* relates to the great body of writing and theory on the nature of human development. While there have been and now are many theories of how children learn (E.Gibson, 1969), perhaps the most familiar and influential work was that of Jean Piaget with his hypothetical stages of knowledge through which children move in their journey to intellectual maturity.

Stages in Human Development

Piaget (1950)[1] described four stages in the development of logical thinking through which a child passes from infancy through adolescence. Each stage represents a new way in which the child/adolescent actively constructs knowledge of the world.

Sensorimotor Stage. In this stage, from birth to approximately two years, children deal with the environment at the perceptual level – they react to it. Their actions take place with little or no understanding. An object exists only if it is touched, seen, or heard; when it is removed from sight or hearing range, it essentially ceases to exist. Only at the end of this stage is object permanence acquired. There is no abstract thinking.

Pre-Operational Stage. In this stage, from two to five or seven years, children are able to remember things, but only by experiencing them. Objects and their relations to one another can be represented through the use of mental images, language and other symbols. The use of symbols depends on children's own perception and intuition, so there is some representational and abstract thinking – hence the term pre-operational. However, the thinking is usually egocentric; children at this level cannot see things from another's viewpoint. They are also very much influenced by visual appearances. No longer are they wholly dependent on what is happening in their immediate environment; they will imitate behaviours that occurred previously. It is an age of curiosity, of questioning, and investigating new things. But since children of this stage may have only a limited experience of the world, they may make up an explanation when experience fails to provide one.

It is in this stage that the children's thinking differs most from that of adults.

Concrete Operational Stage.   From the ages five or seven to eleven or twelve, children develop logical structures to deal with changing objects in the physical world – they are able to perform mental operations. And these operations are reversible. Operations are labeled "concrete" because they are applied only to objects that are physically present. It is a stage which deals with the here and now. Children can speculate on environmental events in such a way that they do not need to experience them through the senses in order to determine the results; events may have been actions previously performed physically. They learn to classify elements of the world, for example, by color, shape, size, or weight. Certain "laws" of logic emerge, such as in the recognition of hierarchies. If, however, something happens in opposition to their logic, they may presume the event is at fault, and not the logic. Conservation, the ability to see that objects or quantities remain the same despite a change in their physical appearance, is considered the major acquisition of this stage.

Formal Operational Stage.   Sometime after year eleven, children are capable of hypothetic-deductive reasoning and abstract thought, of thinking about the future, the abstract, the hypothetical. For example, they can act on two or more abstractions simultaneously in a logical manner so as to come to a resolution about them. Thought is more flexible, rational, and systematic; problems can be examined from a variety of points of view. Children can also think about thoughts and such abstract concepts as time and space.

These are, of course, not watertight compartments. Intelligence and age are important modifiers, and there appear to be inconsistencies in the type of responses children may make in various, but similar situations, and in the responses of one child to the next (Dodwell, 1960, 204). But while a number of investigators have failed to corroborate some of Piaget's findings, Dodwell (1963, 160) makes the useful point that since the sorts of behaviour which Piaget describes as characteristic for certain ages and stages of development have been observed, and since his theoretical account is "satisfyingly coherent," his theory should not be rejected out of hand. Rather, it seems sensible to look for possible reasons why the pattern of development should be blurred. Thus we have no reason not to examine how Piaget's work might relate specifically to education in cartography and geography.

Children's Knowledge of Space

To do this, a more obvious connection can be seen in the work of Piaget with Bärbel Inhelder (1956). They describe three levels of relationships by which children operate in space: topological, projective, and Euclidean.

Topological Space   At this level children come to understand the nature of space through the relations of proximity, separation, order, enclosure, and continuity. These relations are observed empirically and are built up between the various parts of figures or patterns which they organize, independent of any contraction or expansion of these figures. Children at this level, therefore, do not conserve features of those objects or patterns such as, for example, distances, straight lines, angles, and so on. Thus, they view objects in unique experiential relationships rather than in ways that enable them to be related collectively to others (Piaget and Inhelder, 1956, 153).

Projective Space.   At this level it becomes possible to link together or organize different figures relative to one another in accordance with general perspective or projective systems or according to coordinate axes. Straight lines, angles, curves, and distances begin to be conserved throughout transformations.   Thus objects and patterns begin to be viewed in relation to a point of view, either of one's own or of others.

Euclidean Space.   At this level the child is able to apply the concepts of perspective points of view, parallelism, proportions, and coordinates to the solution of general problems within a topographic schema. In other words, all the relationships of Euclidean geometry are preserved.

From the point of view both of mathematical construction and psychological development, Piaget and Inhelder (1956, 301) make it quite clear that both projective and Euclidean space relationships are derived independently, in two different ways, from those of topological space. And while the foundation ideas of space rest upon such Euclidean concepts as straight lines, angles, geometric figures, measurement, and the like, geometrical analysis tends to show that fundamental, spatial concepts are not Euclidean at all, but topological (Piaget and Inhelder, 1956, vii). As they go on to show, a child's active and operational space invariably begins with simple topological types of relationships long before it becomes projective or Euclidean. At first, representational thought appears to ignore metric and perspective relationships and proportions (Piaget and Inhelder, 1956, 4). As a result, the child is forced to reconstruct space from such primitive notions as proximity, separation, order, and enclosure.

Questioning Piaget

While we have accepted for the moment the general value of Piaget's structures, we must also acknowledge the existence of doubters. Hughes, for example, reports on some of those who have re-examined Piaget's findings (1986, 19f). A most thoughtful review can be found in Donaldson (1978). She feels that by the mid-1960s most of Piaget's findings contributed to the same conclusion: that children under the age of seven are not really thinkers; they are very restricted intellectually. But, in fact, other research suggests that we have underestimated the rational powers of children. They are closer to adults than has been supposed, can do more than we have thought, and are not as egocentric as Piaget has claimed (Donaldson, 1978, 58-9). In addition, Donaldson describes a number of experimental findings that, she notes, cannot be reconciled with Piaget's *claims*, but somehow must be reconciled with his *findings* – for these are not suspect (1978, 23). So, for example, Piaget's famous three mountain experiment, which provided him with much of his evidence on children's abilities to decenter, can, in fact, be performed successfully by preschoolers if great care is taken in introducing the experimental problem.

Unfortunately, on just such evidence as this, Piaget's findings have been more rigidly interpreted in geographic education than they needed to be. It should be acknowledged, however, that Piaget did not direct his efforts toward the solutions of educational problems; he described cognitive development, not how it must proceed (Wadsworth, 1971, 120-1). On the other hand, there is other evidence derived from working directly with young children using maps and air photographs. Perhaps the most convincing results of this approach have emerged from the place perception research that was begun at Clark University in the late 1960s and early 1970s.[2] As the researchers at Clark see it: "The problem is straightforward, though by no means simple. So little is known about geographic modes of perception and abstraction – at any stage of human development – that no defensible strategy can be devised for adjusting geographic instruction to the process of geographic learning; for turning our transmitter to the frequency of the receiver" (Blaut and Stea, 1969, 4-5). Thus they set out to "search for coherent insights into the way children acquire geographic concepts through the normal processes of perception and cognition, and the way children abstract from direct, visual perception to the language of maps"

Among their more intriguing findings are the following:

1 that children as young as age three can represent a cognitive map
   (Blaut and Stea, 1974, 5);
2 that in two cultures which have been studied, preliterate children of
   school-entering age can interpret a vertical aerial photograph, abstract from the photographic image to a system of highly iconic map
   signs, and use the reduced, rotated, abstract presentation in solving
   a simulated route-planning problem (Blaut, McCleary, and Blaut,
   1970, 346);
3 that a third grade child can interpret the major categories of macro-
   scale features in an urban environment, maintain interest with unfamiliar environments presented at various scales, and abstract features representing major land uses and place wholes (Hart, 1971,
   63-4).

These kinds of findings suggest that preliterate children of school-
entering age can engage in a real, if primitive, form of map reading, map
making, and map use (Blaut, McCleary, and Blaut, 1970, 346). They can
deal with map-like representations and display immense pleasure in
doing so (Blaut and Stea, 1971, 392), and, by third grade, "can interpret
the major categories of macro-scale features in an urban environment"
(Hart, 1971, 6-64).

Another problem that has arisen with the four-stage Piagetian model
is the implication that the stage of formal operations is the end product
of the maturation process. We now see, however, a number of studies
(Common, Richards, and Armon,1984) that are looking at intellectual
developments beyond the formal operations, a stage which is called
"the general systems thinking world." Since this stage relates to development beyond age sixteen, it does not pertain directly to our discussions here. However, it may be shown that some of the thinking skills
at this new level are directly related to some of the earlier experiences
in geographic thinking.

Implications for Education

It would appear then that Piaget has provided a rich developmental
picture from which there should be plenty of possibilities for developing
varied approaches to geographic problems and concepts. In addition,
subsequent research has shown that any use of Piaget's stages of development to limit children's exposure to maps was unwarranted. However,
such a conclusion must be conditional upon our assurance that children
understand exactly what is asked of them and that the questions and
activities offered are in themselves related to the stages of development
and to the appropriate kind of "space" in which they are "operating."

WHAT CHILDREN ARE TAUGHT

This is an equally difficult question to address for it is impossible to document even a reasonably large sample of all that takes place in classrooms across North America. However, from an examination of textbooks, curriculum guidelines, papers in professional journals, and teaching materials handed out at meetings of professional educators, it is possible to infer something about what is taught. Four themes seem recurrent.

First, we tend to introduce children to maps and mapping through some very sophisticated and abstract concepts without, apparently, making sure some of the prerequisite skills and concepts are in place or have at least been introduced. Second, there is great difficulty in distilling the wealth of educational and psychological "research" on map learning because of the great disparity of contexts, tasks, and terminology utilized in these reports. Third, whatever has been done in classrooms has not been very effective for there is ample evidence illustrating the geographic illiteracy of North Americans and the decline of geography as an academic subject in schools. Fourth, for all the attention which Piaget's work has attracted, there is little evidence from teaching materials of its positive impact upon the structure of geographic curriculum in the earliest grades.

There is firm evidence to support these conclusions. First, many geography curriculum guidelines contain statements that relate either directly to a very abstract concept or state the concept with sufficient generality that either (a) the underlying concepts or skills are not identified or (b) the abstract aspects are stated in such a way as to be easily translated into key words or the structural elements of maps for memorization by students. For example, there are aspects of both these possibilities in the simple curriculum directive "knows and uses cardinal directions" (AAG and NCGE, 1984, 12). The use of cardinal directions is not as simple as it may appear, for they are part of an absolute, and thus abstract, system for specifying direction. In contrast, the assignment of directional labels (north, south, east, west) to various walls of a classroom is not particularly difficult. But it is quite a different matter to understand that these labels transcend the classroom and bear abstract and complex relationships to the sun, Polaris, and the axis of the earth.

Thus the process of orienting oneself in a real world context is the true underlying skill that should be enunciated and mastered before a more abstract system of orientation is introduced. What is missing then, are the various procedures by which students learn to orient themselves to more confined and restricted areas using visible landmarks in a relative way. Unfortunately, many lessons in orientation, found in various printed materials, emphasize the prominent features of this abstract system which are easy to memorize – parallels, meridians, the direction names, and so on. In contrast, little is done to introduce the basic skills and concepts of orientation that operate in personal space and perhaps in small-scale thematic maps as well.

Commendably, the guidelines noted above prefaced the item on "knows and uses cardinal directions" with three other suggested learning outcomes:

1 Knows and uses terms related to location, direction and distance (up/down, left/right, here/there, near/far).
2 Knows geographic location of home in relation to school and neighborhood.
3 Follows and gives verbal directions (here/there, left/right).

Achievements in these activities are undoubtedly appropriate for the kindergarten through second grades for which they are suggested and for the micro-areas to which they are most applicable. In addition, these achievements lay the groundwork for later orientation activities using cardinal directions. Obviously cardinal directions are eventually useful at all geographic scales, but only after the underlying concept of orientation has been developed progressively in the classroom from the local to the global situation.

Second, there is a vast literature on teaching geography at all school levels. What makes it so difficult to evaluate this body of work is that the specific tasks involved are not always clearly spelled out. Later on, in chapter 7, I will make a clearer distinction between the terms "activity" and "task" as they relate to map use. For the moment, we can consider an activity as something that is done – a measurement is made or a name searched for and found. Activities closely relate to goals: What do I want to measure or look up? Thus they are tied directly to the kinds of information that should be present. Tasks, on the other hand, are quite specific operations of the eye-brain that are independent of the kinds of information involved. In the context of map use, they are per-ceptual-cognitive operations performed on the graphic image.

To clarify this distinction, it has been noted (Meyer, 1973,29) that primary grade children cannot be expected to accurately *relate* (my italics) distance on a map to distance in the world by the use of scale or proportions. In this context, measuring distance on a map is an activ-ity. It involves the task of matching two points, separated by some distance, to markings on a measuring device, whether they are drawn with a pencil on a blank piece of paper or marks on a ruler. The hand-

eye coordination of manipulating the ruler and the alignment of the
points on the map with marks on the ruler (and perhaps estimating the
fraction between two ruler marks) are all visual tasks. It is upon the
mastery of these tasks that skill in the activity of distance measurement
rests.[3] The use of a ruler to formally and consistently describe the
distance between two points is probably one of our most basic analytical skills. In Ontario, it is a part of the Primary Division Curriculum
(Ministry of Education, 1975), for mastery by the end of the third grade.
In contrast, *relating* distance on a map proportionally to distance on the
earth is a much more complex and abstract notion, and one that likely
cannot be "taught," only experienced. Our best hope is that we can
devise a suitable way to provide that "experience."

Third, whatever we can determine has been done in schools, it is clear
that it has not been very successful. Recently, the media has made a
disturbing number of references to the deplorable state of education in
general and geographic knowledge in particular. For example, a report
by the eighteen-member National Commission on Excellence in Education (*Time*,1983) warned that:

Our nation is at risk. The educational foundations of our society are presently
being eroded by a rising tide of mediocrity. If an unfriendly foreign power had
attempted to impose on America the mediocre educational performance that
exists today, we might well have viewed it as an act of war. We have, in effect,
been committing an act of unthinking, unilateral educational disarmament.

The report went on to declare that geography teaching needs improvement, that high school geography courses, although offered, were
completed by only 16% of students in a recent sample of high school
graduates, and that American students spend inadequate amounts of
school time in geography instruction when compared to other industrialized nations (Eric, 1985). The report also revealed such estimates as:
40% of minority youth are considered functionally illiterate; from 1963
to 1980, the average scores on Scholastic Aptitude Tests fell more than
fifty points in verbal skills and thirty-six points in math[4]; and there was
a pronounced rush from tougher to easier or "more relevant" subjects.
For example, the percentage of highschool graduates who took English
dropped from 96.9% to 76.5%; the percentage taking algebra decreased
from 75.5% to 63.8%; those taking western civilization from 87.6% to
41.7%. In contrast, the percentage taking general social studies increased from 8.0% to 22.0%, the percentage in health and physical
education increased from 0.8% to 12.0%, and the percentage in driver
education increased from 0.3% to 58.6%. Clearly, there was a switch
away from academic subjects in favor of "practical" subjects. It is

difficult to determine from these figures just how well geography fared since one would expect there to be a significant component of geography in both the study of western civilizations and social studies.

In 1982, the US Department of Education estimated that less than 9% of all secondary students in the country were enrolled in geography courses – an enrolment drop of 40% since 1960-1. As one journalist noted recently, "only in crisis do Americans decide to learn geography."

During the same period, over two thousand students in introductory college classes at the University of North Carolina were surveyed and tested on geography. The percentage of students never having had a geography course was high – 71% never had geography instruction in elementary grades, 65% in junior high, and 73% in highschool. Only 12% were able to name the Great Lakes, as compared to a similar survey in 1950 when 46% could. Of the thirty countries in Africa south of the Sahara, only 8% of the North Carolina students could name *three*; 69% could not name one!

Perhaps the most discouraging statistic came from a review of a number of studies illustrating the level of geographic illiteracy. Geography tests, dealing with questions about relationships rather than place name items, were administered to twelve-year-olds in eight industrialized developed nations. American students ranked a weak fourth among the eight groups. Twenty percent of one group taking the test could not locate the United States on a world map (Eric, 1985)! Clearly, the report concludes, "geographic knowledge of high school students is inadequate and ... enrollment and achievement in geography education are low."

Fourth, what is interesting about the stages of intellectual development described by Piaget is that so few of them are reflected in the educational materials suggested for introducing maps and geography in schools. This is especially curious in light of the great regard in which he is held by educators. For example, children first come to understand the nature of space through various topological relationships. These relationships are described in any simple linkage diagram which shows, for example, the successive stops on a bus or street-car line. Children's appreciation of scale is imperfect, but not their ability to deal with such qualities as proximity, order, continuity, or connectivity. Thus the representation of topologic relations among objects or points would seem to be a useful type of mapping in which to engage. Unfortunately, such "topologic maps" are rarely suggested or promoted in elementary teaching materials even though they match both the theoretical level of the children's development and, undoubtedly, their personal experiences of space. The rarity of their use in school stands in sharp contrast to their wide use in the adult world.

In contrast, lessons in early grades often deal with parallels and meridians, cardinal directions, and map projections. These topics relate to some of the most abstract expressions of the concepts of location, direction, and orientation on the surface of the earth, and to distances between points on various transformations of the spherical surface to a plane! Perhaps the very ease with which teachers can reduce these concepts to memorizing key words or component elements of their expression in maps has made them so popular. Teachers should not be deluded, however, into thinking they are teaching the understanding of these abstract ideas.

As applied formally to the practice of teaching about maps and mapping, numerous attempts have been made to produce graded lists of the skills that are appropriate to each Piagetian stage or level. For example, Rushdooney (1968), in summarizing over eighty pieces of research on children's abilities to read maps, derived a ten-step gradation of "map-reading skills" which "children tend to know and can learn when systematic instruction is provided at their maturity level" (Rushdooney, 1968, 213).

This sort of exercise should in itself be beneficial. However, it seems to have led to an unfortunate emphasis on the limited capabilities of children at each stage, rather than on what they could do. There also seems to have been an unfortunate shift away from developing intellectual operations in or about space, which were the subjects of Piaget's classic experiments, and toward teaching about attributes of maps themselves. For example, Rushdoony's ten-graded ability steps are subdivided under six topics: size and shape, orientation and direction, location, distance, symbols, and map inferences. While it is difficult to know exactly what was meant by some of the entries, they are clearly progressive in difficulty. In the case of distance, for example, relative judgments of near and far are noted in preschool, the use of street blocks in grade two, and only in grade five is the use of a scale in miles mentioned. But since a scale bar appears without explanation on most maps available to elementary school children, it may be easier to study it and use it than to systematically examine the ways we might measure and represent distances in space, in both relative and absolute terms.

Perhaps a more useful statement was made by Fraser and West (1961) who stated that an individual who has an adequate command of geographic understanding and skills exhibits at least six related characteristics: (1) a sense of direction, (2) a sense of distance and area, (3) an understanding and use of geographic vocabulary, (4) an ability to obtain information from maps and globes, (5) an ability to visualize continents and regions of the world, and (6) an understanding of man-earth relationships.

Certainly having a "sense of," for example, distance is ambiguous as a curriculum item unless we view it as the unifying concept over a number of years in progressively (1) using words like "near and far," (2) pacing distances, (3) comparing graphically the length of an object with its length in a drawing, (4) using a measuring tape, (5) using scale bars or other designated distances (as between two familiar points) on large-scale maps, and (6) calculating arc distances between points on the spherical earth.

Jarolimek (1963, 20), in examining skill development in social studies, makes this point by saying:

To help pupils develop a skill to an advanced level, the component elements must be identified and arranged in a sequence representing levels of difficulty. The learner is then introduced to these components one step at a time in a developmental pattern. These component elements should in themselves be functional in that the learner can apply them to social studies experiences at his present stage of development. That is to say, skills should not be broken down into fragmented elements which serve no immediately useful purpose.

In other words, the arrival at a sense of understanding about some concept will rest upon experiencing it in different ways and at increasing levels of difficulty.

The breakdown of a true skill into component elements that are in themselves functional, and that the learner can apply to social studies experiences (to paraphrase Jarolimek), is more easily accomplished with unifying concepts or processing operations than with specific map elements. For example, we do teach and work with pictographic and conventional symbols *first*, and then later introduce more complicated symbols for use in thematic maps or in specialized map products such as hydrographic and aeronautical charts. Rarely do we accompany these maps with instructions as to the logic of their construction. In a sample of twenty-four twentieth-century North American children's atlases (Castner, 1984b, fig.7), thirteen provided a legend of conventional symbols, but only one gave any information on how symbols work, that is, how their graphic dimensions are made to convey their intellectual dimensions.

But what is more disturbing is that in providing map symbols without this sort of explanation, we are avoiding the component problems of symbolization. Yet, if we allow children to design their own symbols, then we are actively involving them in observing, representing, scaling, generalizing, and abstracting (without naming them as such) information about objects around them. Even with the freedom to make and use their own symbols, in all probability pictorial symbols will be among the first they will create. Only later will they need to consider the problems of symbol design in the context of the international and intercultural requirement for agreement on common signs. At this point, they may more readily understand why some symbols have become conventional.

As it is, we can only suppose that atlas makers and curriculum writers have assumed that information on the logic of symbols has been or is being taught elsewhere. But it is very difficult to find evidence of this among published and informally distributed teaching materials on map skills. The same sort of arguments that we have made here for teaching about symbols could also be made for the other topics in Rushdoony's list.

There is a disturbing aspect about lists of map skills – the implicit independence of each topic. Yet it is very difficult to think about orientation and direction, for example, without a location or locations in mind; to consider size and shape and not symbols. Thinking about some aspect of the earth's surface, and the problems of representing or describing it is much like playing bridge. One cannot play that card game without knowing all the rules; but most of the rules will not be understood unless one has played the game. This is a Catch-22 situation! But of greater relevance is the fact that children have already been thinking about and evaluating their environments for years before they arrive in our classrooms. They may even have drawn crude pictures or sketches in an attempt to organize and express their thoughts. Unfortunately, their drawings are probably judged by their failure to resemble adult maps, rather than by their successes in providing useful spatial information, as primitive as we may consider their attempts.

Turning Off Students

One explanation of the state of geographic knowledge may be that there is something about the way children are introduced to geography that turns them off the subject. This may mean how they have been introduced to maps. The following analogy suggests an alternate way in which children might be introduced to maps.

Introducing geography to children through the study of the conventional and structural elements of maps, outside of a geographic problem-solving context, can be seen to parallel the problem of helping disadvantaged children read once they have entered school. Mary Howarth is an elementary school reading specialist who works with disadvantaged children. She presents compelling evidence that traditional reading instruction does not build on language usage that the child brings to school. She also suggests that building a child's self-concept is a basic key to the mastery of the reading process.

For many disadvantaged children, "ain't" is a perfectly normal word. "He don't" is what Daddy says; "Youse" is how Mommy expresses the plural when she addresses more than one person. In other words, many children do not share the standard middle class English patterns and words that are being taught in schools. For children who are learning English as a second language, *patois* is the only English they know (Howarth, 1984, 2). For these kinds of children, a good prereading program is built around oral language and would begin with the teacher becoming familiar with the children's language patterns. At the same time, the teacher must expose them to the "teacher's language" patterns and create an interesting environment where there is a need to *communicate and there are things to communicate about* (my italics) (Howarth, 1984, 3). The teacher's critical role is described in the following way: "Although the teacher does not correct a child's speech at this stage, neither does she make any attempt to accommodate to it by altering her own speech patterns. These children need to hear correct language usage over and over, but there is no point in confusing the teaching of reading with the teaching of standard English. Bridging the gap between the child's language and standard English is built into the programme in gradual, non-threatening stages" (Howarth, 1984, 3).

The prereading activities are carefully chosen to include games, having the children illustrate something about their stories, printing words, and reading for enjoyment only. The activities are designed to compensate for voids in the children's experience and to ensure their success at every stage in the process. The learning of words is always done in context (in sentences, never in isolation). As the children progress, they move from their own individual material to that created by someone else until the gap is bridged.

A similar contrast has been drawn between "spontaneous art" and "school art" (Wilson and Wilson, 1982, xv-xvi). The former comes from children's own desire to present and experiment with developing ideas about themselves and the world. With spontaneous art they can express, review, and re-examine their own realities more easily than with written words and more fully than with numbers. Unfortunately, such art is usually less colorful and visually compelling and thus meets few adult expectations. In contrast, school art is seen as being of greater educational value because it meets adult conceptions of what children's art ought to be!

All this suggests that we would be more successful in introducing geographic ideas and concepts to children if we allowed them to utilize their natural inclinations in communication rather than the conventional products of the adult world. Blaut and Stea (1971, 393) express a similar thought by noting that geographic learning begins very early and develops in a manner that is discoverable, if still largely undiscovered. If we knew more about the informal or natural modes of learning in the elementary school years, we would be in a better position to design curricula that would build upon these modes, rather than smother them. Gradually, as children grow in the understanding of the innate processes they employ in learning, they can go on to more sophisticated activities and work up to the standard, conventional procedures we use as adults.

Implications For Geographic Education

It would appear then that there is some variance between what we considered was relevant in Piaget's theory, and what is in fact possible. Thus we may have been too restrictive in our treatment of mapping concepts. It is as though we compared Piaget's developmental stages to the attributes of formal, Euclidean models of maps and found a match only in the formal operational stage. As a result, we assumed that there could be no map work before this stage. In so doing, we have failed to consider what other kinds of "maps" and graphics match the skills of children in the pre-operational and concrete operational stages. As a result, we have not developed any formalized mapping program for children who are deemed to be in either the pre-operational or concrete operational stages, that is, children who are operating in either topological or concrete space.

WHAT CHILDREN SHOULD BE TAUGHT

If this is a value-laden question (Petchenik, 1984, 14), then there should be some logical basis for developing some answers. Two possibilities that have great appeal here are activities that utilize (1) critical thinking skills and (2) behaviours observed by Piaget in the various stages of intellectual development, particularly those in the pre-operational and concrete operational stages and in topological space.

Critical Thinking

There is no general agreement on what constitutes the critical thinking skills. Beyer, who attempted to provide the groundwork for a definitive statement on critical thinking did so by first noting what it is *not* (Beyer, 1985). Then, by reviewing the many attempts at its definition, he was able to provide a three-part conceptualization of what any critical thinking skill should incorporate.

First, we should examine what critical thinking is not. The skills of critical thinking should be separated from the skills of problem solving and decision making. If this is not done, we might fail to distinguish the unique features and functions of critical thinking from these other forms of thinking. For example, gathering information or "grasping the meaning of a statement," are not unique attributes of critical thinking (Beyer, 1985, 274). In contrast, many of the large number of skills enumerated for social studies (Johns and Fraser, 1964) include few of what Beyer would regard as critical thinking skills. For example, using a table of contents or recording information from a field trip would not qualify as critical thinking skills.

Second, we should examine what critical thinking must be. Beyer's review provides a number of statements by various writers commenting upon the nature of critical thinking. Among the more compelling in light of our interest in improvization and mapping are the following:

Critical thinking is the process of examining ... materials in the light of related objective evidence, comparing the object or statement with some norm or standard, and concluding or acting upon the judgement then made. (Russell,1956,285)

Critical thinking...consists essentially of 'evaluating statements, arguments and experiences'(D'Angelo,1971,7). It involves in its most benign form 'the correct assessing of statements' (Ennis,1962,82-3), and in its most aggressive form 'the spotting of faults.' (deBono,1983,706)

It is both a frame of mind and a number of specific mental operations. (McPeck,1981,162)

It involves 1) an *alertness* to the need to evaluate information, 2) a *willingness* to test opinions and 3) a *desire* to consider all viewpoints. (Fraser and West, 1961, 222)

In other words, critical thinking is the assessing of the authenticity, accuracy, and/or worth of knowledge claims and arguments (Beyer, 1985, 271). Beyer's (1985, 272) composite list of the core analytical and evaluative operations that are included in critical thinking are:

• Distinguishing between verifiable facts and value claims.
• Determining the reliability of a source.
• Determining the factual accuracy of a statement.
• Distinguishing relevant from irrelevant information, claims, or reasons.

- Detecting bias.
- Identifying unstated assumptions.
- Identifying ambiguous or equivocal claims or arguments.
- Recognizing logical inconsistencies or fallacies in a line of reasoning.
- Distinguishing between warranted or unwarranted claims.
- Determining the strength of an argument.

Third, we should consider what critical thinking should be. More specifically, Beyer proposes that for any individual critical thinking skill, there are three essential kinds of attributes: (1) a set of procedures by which it becomes operational; (2) certain distinguishing criteria or clues that serve as evidence as to what it is, for which we search by using the set of procedures; and (3) a set of rules which tell the user when to employ the skill, what to do when certain clues cannot be found, what to look for to identify revealing patterns, and in general to make the skill workable.

Beyer goes on to describe in somewhat more detail what he means by procedures, criteria, and rules. Procedures are the steps or operations that one employs in executing the critical skill. For example, in searching for bias in a statement, we would look for emotionally charged words, slanted data, overgeneralizations, or rhetorical questions. Finding these words, particularly in a clear and persistent pattern, provides the criteria by which we can judge the statement to be biased. The rules provide guidelines about searching for clues, what to do when certain clues cannot be found, how to look for revealing patterns, and what to look for in various combinations of clues. Thus, in the above example, the search for "loaded" words or statements is a rule. Other rules come into play to make the skill workable under other conditions, for example, when the statement is in another *medium*.

But when all is said and done, the teaching of critical thinking skills appears to have an agenda of its own. For example, Beyer notes that the "standards that distinguish each critical thinking skill must be taught and learned as knowledge." And while there is little to be found in the literature about the rules, articulating them is an important key to mastering the skills (Beyer, 1985, 276).

Thus, while there is a lot of speculation on just what critical thinking skills are, there are few concrete statements of the procedures, criteria, and rules for specific critical thinking skills. This is apparent in the area of map skills because, by and large, the curriculum recognizes such a narrow range of map purposes. Map making is considered to be a rather singular process of recording the location of objects or places – an inventory process. As a result, the intellectual richness of distinguishing criteria, clues, and rules for the employment of mapping skills have not been given to students.

The absence of specific statements on critical thinking skills in carto-graphic expression is undoubtedly due to the relative youth of formal research in cartographic communication. There has been a critical mass of research scholars in the discipline for, perhaps, only two decades. This does not mean that opportunities have not existed. Almost every time we have undertaken to produce an historical atlas, we have had to contend with problems of missing data, unreliable data, uncertainty about boundaries and locations, conjecture and hypothesis about events, influences, and relationships. Yet rarely have we set out from the beginning to provide alternative graphic solutions in our designs that would make it possible, if not obvious, to the subsequent atlas readers which of these problems was present and how it was handled. In all likelihood, a "considered opinion" was rendered, and a normal design solution was adopted that buried forever the hard decisions and compromises that were made.[5]

I note this particular practice with historical atlases for it seems to provide one vehicle whereby the objects of many critical thinking skills can be tested. Certainly in many of the cartographic applications of the process of generalization we have the opportunity of creating, examin-ing, and judging the results of decisions made in leaving out various kinds and amounts of information. By then turning the process around, we can describe the kinds of critical thinking skills we believe to be applicable to these images and the information they represent. Only then can we then begin to generate the procedures, criteria, and rules that will apply to critical thinking skills in cartographic statements on geographic problems.

Behaviours Observed by Piaget

A second attractive approach to the question of what children should be taught is to make more positive use of the behaviours which Piaget does associate with his various stages of cognitive development, par-ticularly those in the pre-operational stage, the concrete operational stage, and in topological space.

I have suggested previously that the abilities which Piaget associates with each stage may have been interpreted too rigidly or perhaps too negatively, namely in terms of what map related activities might take place at a given level. If students are unable to perform a given operation at one stage, then, the argument goes, we must put off efforts at teaching the concept to which the operation is related until the child reaches a higher stage of development. This argument, however, may tend to overlook what children can do.

For example, students in the pre-operational stage are not able to think from another's point of view. Thus an architectural plan of an unknown building interior will not be very meaningful. But the research noted above finds no difficulty with preschool children looking at aerial photographs and at areas familiar to them. Obviously they are experiencing a scene from a vantage point other than the one they normally have. Experiencing multiple views of any object or scene would seem to logically precede being able to *imagine* the scene from other points of view without having the pictures or drawings of those views present. I have not found a specific piece of research establishing this, but it seems intuitively logical.

In the concrete operational stage, the child is able to classify elements of the world by various criteria and can recognize hierarchical structures. Certainly there is nothing more structured in these ways than the *information* found on a standard topographic map. While understanding the information found on this type of map involves acknowledging the meaning of its symbols, it is not necessary to understand the logic of the symbols' design – that is an important but separate study. What I am suggesting is that, for example, examining the various ways in which the map provides information on how people can move about (transportation modes or routes) is inherently as simple as understanding how canals, railroads, airports, and roads of several types are symbolized. This would be particularly true of concept-related or associative and abstract symbols.

Take, as an example of an abstract symbol found on topographical maps, the lines connecting points of equal elevation above some reference surface (contours). They are a much more complex phenomenon than the information that they reveal – how high it is in a particular place. Or, the case of the solid green vegetation symbol which depicts areas where a compiler has judged that more than some threshold percentage of the area is covered with trees. Such symbols may therefore cover areas with quite different densities of trees (and even some open areas) and may exclude other areas with trees.

Implications for Geographic Education

The question arises as to how critical thinking skills might be incorporated into a geographic educational scheme that also takes advantage of the structure of Piaget's ideas. One way is by addressing concepts through questions which have more than one answer. This makes it possible to examine and evaluate various clues, arguments, and criteria rather than to focus on memorizing place names and terminology. This method would seem to be advantageous for teachers not confident in their geographical training. Their role is changed: instead of feeling they must know all possible, up-to-date answers before beginning a lesson, they become judges and adjudicators. This new role requires them to decide what facts are admissible, what arguments are valid and logical in the face of the criteria chosen to evaluate the various solutions. These, of course, are skills teachers bring to all of their classroom activities. With a variety of possible answers, more class members can also become active participants.

This atmosphere, in turn, makes it easier for questions to be posed by the students. That this method makes learning much more meaningful is so obvious that it should not be belaboured here. However, it is important that the context of any activity be easily related to by the student. This should condition any attempts to create teaching situations that provide for alternate answers and the utilization of critical thinking skills.

CONCLUSIONS

The identification of geographic concepts which can be manifest at all developmental levels will enable us to teach them in sequences of increasing complexity and sophistication. For example, consider Piaget's description of the sequence with which children can make judgments about problems with number. At first, children are able only to comment on what they *perceive*, (a "global" estimate of some possibly irrelevant stimulus dimension), and this from a very egocentric perspective. Estimates of the number of beads on a string, for instance, will be modified by changes in the way the beads are arranged. This is followed by an "intuitive" stage in which children begin to realize that quantity or number of beads is an attribute which remains invariant despite perceptual transformations, (changes in the shape of the string). Only when children's judgments become completely "operational," consistent, and reversible will they be fully conversant with the concept of number (Dodwell, 1960, 192-3). Yet at all these levels, children are considering the concept of number; what is changing is their ability to deal with the critical dimensions of the string of beads, (to differentiate its changing and invariant aspects).

Piaget was quite clear in his belief that the student must be an active part of the educational process. He must transform things and find the structure of his own actions on them (1954, 4). It is this cognitive reorganization, made available by "self-discovery" in the classroom, that Piaget, Dewey, and Montessori all stress as a crucial element in child development (Hooper, 1968, 429). In this process, it will be the teacher's task to analyse the content to be learned in terms of the operations implicit upon it. "Having done this, he will arrange the learning materials so that these operations can actually be carried out by the student himself, and then see to it that the student does carry them out" (Flavell, 1963, 368).

The task set for children must be just beyond their reach so that the right amount of discrepancy is created between their established schema and the discordant input. The learning takes place when children are able to resolve the conflict through an adjustment of their own schematas, that is, they discover the similarities between a new situation and a slightly different one known before. Piaget calls this mechanism the "equilibrium process."

In a similar way, Jerome Bruner (1966, 44-5) also described a progressive set of actions (which can utilize images, pictures, and symbols) that can bring about the acquisition of ideas. The most primitive, which he calls *enactive representation*, involves children's contact with their environment by means of movements through it and contact with objects in it. These contacts are translated into a cognitive set of spatial associations. Learning, then, would take place only when the child interacted with real objects in the environment – a point also made by Piaget (Hooper, 1968). In contrast, *iconic representation* is greatly dependent upon the various perceptions that children experience within their "world." As such, it is relatively independent of action. Learning here would make use of various kinds of realistic models. The *enactive* and *iconic* modes of learning lead developmentally into *symbolic representation* in which signs are used systematically according to certain rules. The most specialized form of symbolic activity is, of course, the development of language.

Bruner recognized that it was not necessary that each mode be experienced in turn. Rather, a child could learn by some combination of the three modes (1966, 33). Putting it another way, there are various levels to any concept, and there is one appropriate form with which the concept can be introduced at any given age.

What is needed in cartographic and geographic education are ways in which our basic concepts can be introduced in each of Piaget's stages and Bruner's modes. The remainder of this volume is one attempt to determine what concepts might be most appropriate to accomplish this. Since most of our assessments of the world will be made through the agency of vision, we should consider next the basic mechanisms of seeing and visual perception.

CHAPTER THREE

The Visual System

A literary narrative or a piece of music unfolds in stages, but in a still-life time is fore-shortened as it were, and by taking it in with a single sweep of the eyes (or so it seems) this multitude of experiences blends into one near-simultaneous process, so that it is extremely difficult to sort out the various elements which went into its making. The trouble with explaining visual beauty, and also its fascination, is that so much is happening at the same time.

Arthur Koestler (1964, 371)

I argue that the human brain has a modular-type organization. By modularity I mean that the brain is organized into relatively independent functioning units that work in parallel ... frequently ... apart from our conscious verbal selves. That does not mean they are "unconscious" or "preconscious" processes and outside our capacity to isolate and understand them. Rather, they are processes going on in parallel to our conscious thought, and contributing to our conscious structure in identifiable ways.

Michael Gazzaniga (1985, 4)

look (look), v.t. 1. to observe, inspect, examine. 2. to search for. 3. to direct one's gaze to in order to perceive or find out; it implies volition.

One of the central themes of this volume is that looking with insight and discrimination is a basic perceptual skill, of equal value in life as skills in listening, in language, and in mathematics. Hence separate chapters are required on both the mechanics of looking (chapter 3) and on looking in action (art and drawing, chapter 5).

Cartographers have long been interested in the former topic in an effort to better understand the perception of maps and to determine the limits within which their alterations in map design could intercede in

Figure 1 An outdoor scene.

that perception. Thus the first part of this chapter is devoted to a review of some of that effort and to some of the findings that will provide in chapter 7 some design leverage on the perception of maps. Central to this discussion will be the differentiation between spontaneous viewing and task-specific looking.

The latter part of this chapter attempts to articulate the attributes of images that carry the information that is actually perceived and then interpreted. The visual analysis of an image involves the discrimination of contrasts among elements and of structures within these contrasts. If we are to utilize maps and other complex graphics for activities other than merely recording and looking up data (as in an encyclopedia), then we must go beyond the symbols we create to examine the relationships between and among them. In addition, we must come to consider mapping as an active, goal-directed process of thinking and drawing.

Some use will be made of the work of James J. Gibson, among others, to provide evidence of the kinds of image characteristics which reveal these more subtle relationships. After all, it is the education of our visual skills which will allow our looking to become more discriminating, both of contrasts and of structures buried in complex graphic arrays. Perhaps teaching more about the existence of these kinds of image characteristics is a way of breaking away from the "here-is" approach we have heretofore taken in early geographic training.

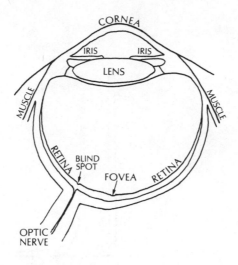

Figure 2 Schematic diagram of the eye.

VISION

The visual perception of the world around us, as in the scene in figure 1, depends upon the interaction of four elements:

1 Light energy from the sun.
2 The substances or materials which reflect that energy and thus determine its quality and make-up.
3 The eye, the receptor of that energy.
4 The brain which interprets that energy.

Light Energy

The visible light pouring to us from the sun is only a fraction of the much broader range of radiation that includes, at the shorter wave lengths, X-rays and ultraviolet light to infra-red and radio waves at the longer wave lengths. The small range of wave lengths that make up what we refer to as the visible spectrum contains all of the colors of the rainbow.

Surface Materials

Our visual perception of the world comes from our interpretation of the light reaching our eyes. This light, originating from the sun, has been reflected from the various surfaces and materials that make up our

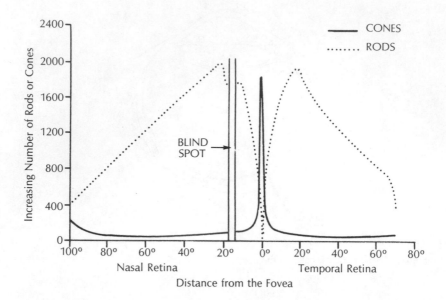

Figure 3 Cross section of the retina showing the relative distribution of rods and cones. After Schiffman (1976, 164).

surrounding environments. It is the discrimination of the relative rate and quality of these reflections that allows us to learn about the nature and position of these surfaces and materials.

The Eye and the Brain

The structural components of the eye can be compared simplistically with those of a camera. There is a lens, with an iris (diaphram) which controls the amount of light that passes through the pupil (shutter). Through the contraction of various muscles an image is brought to focus on the retina (film) where the image is recorded and subsequently processed (figure 2).

The heart of the system is the retina which contains some 126,000,000 receptors of two kinds: rods, which number some 120 million, and cones, which make up the other 6 million (figure 3). Riggs (1971, 282) is less precise, estimating that there are more than 100 million rods and something less than 10 million cones.

Two general distinctions should be made between rods and cones. The first, and most obvious, is their almost reciprocal distribution which is pronounced in an area known as the fovea. The second is that

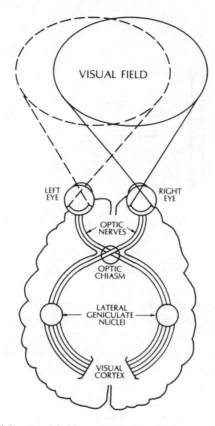

Figure 4 Diagram of the visual fields transmitted to the brain. After Albarn and Smith (1977, 39) and Samuels and Samuels (1975, 56).

cones tend to have individual linkages to the brain, while the rods tend to share linkages. Using the telephone company analogy, cones have direct, private lines, while rods must share party lines. For example, it has been estimated that in the middle and far periphery of the retina, some 100 rods converge on 17 diffuse bipolars, which in turn converge on a single ganglion cell (Brown, 1965, 51). For this reason, the fovea, the area of the highest density of cones, is the area of greatest visual acuity. There resolutions measured in seconds of arc are possible, depending on the particular acuity measure (see Riggs, 1971); in peripheral cells, acuity is in the order of three degrees of visual arc. The fovea itself is an area that subtends about five degrees of visual arc; this translates into about one centimeter at normal reading distance.

Signals from all receptors are carried out of the eyeball at one point, the blind spot, through the optic nerve, which contains some one

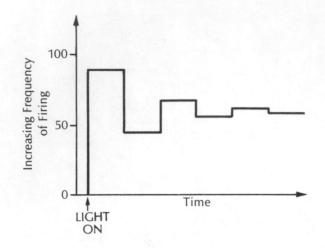

Figure 5 A damped oscillation due to some inhibition effect. After Cornsweet (1969, 19).

million fibres (Brown, 1965, 51). The optic nerves from the right and left eyes come together at the optic chiasma, where fibres from the nasal halves of both retinas cross and join the fibres from the temporal retina of the opposite eye (Graham, 1965, 52). The fibres in the optic tracks extend on to the lateral geniculate nucleus and are distributed to the brain in various ways that we are only beginning to understand. What is significant here is that both halves of the cortex receive signals and are able to communicate through a two-way connection, the corpus callosum. As a result, there is no single place in the brain which is responsible for vision. Rather there are various places where visual information may be processed in parallel (Gazzaniga, 1985, 22-39). Consequently, there is more than one way in which the brain may be able to deal with a visual problem. Figure 4 suggests the nature of this distribution system.

The basic response mechanism of a retinal receptor to the incidence of light is some kind of neural discharge. Some of it is directed to the brain, some of it to its neighboring receptors. A receptor, in turn, will receive a certain amount of feedback from its neighboring cells depending upon how they are stimulated. The result of this kind of feedback is an oscillating reduction in the level of the given receptor's neural output. The rate of this oscillation depends on the distance between the receptors: the further away, the slower the oscillation. The final output level will depend upon the strength of the inhibition effect, and how many receptors are involved: the more linked together, the lower the final level (figure 5).

Any visual system, like that of humans, which exhibits a recurrent, inhibition effect will show a strong output when stimulated and will then drop off rapidly. In other words, such systems that exhibit a recurrent, inhibition effect transmit information about *changes in intensity*, while suppressing information about steady levels of intensity. What excites the eye is not an even distribution of light but at least one contrast or edge; not a fixed level of intensity but a change (Gibson, 1965, 67). The analogy can be made by thrusting one's hand into a bucket of hot water; the sensation of uncomfortable heat soon passes even though the temperature of the water has not changed appreciably. Obviously the danger of trying this experiment in very hot water is that your hand may become burned before you realize it. The reader can experience this by examining the three illustrations in figure 6. The contrast between these responses is shown schematically in figure 7.

The same effect can be seen at the boundaries between the bands of gray in a gray scale. The simultaneous contrast of two unlike intensities along a common border produces an enhancement or lightening of the lighter edge and a depression or darkening of the darker edge. This produces the distinctive scalloped or undulating image; an effect shown schematically in figure 8.

GRADUAL BOUNDARY

Figure 6 Detecting circular boundaries of varying definition. The viewer should focus to the right of each circular area and notice the difficulty or ease of detecting in peripheral vision the circular boundary when the boundary is a) gradual or blurred, b) abrupt, and c) contoured. After Cornsweet (1969, 21).

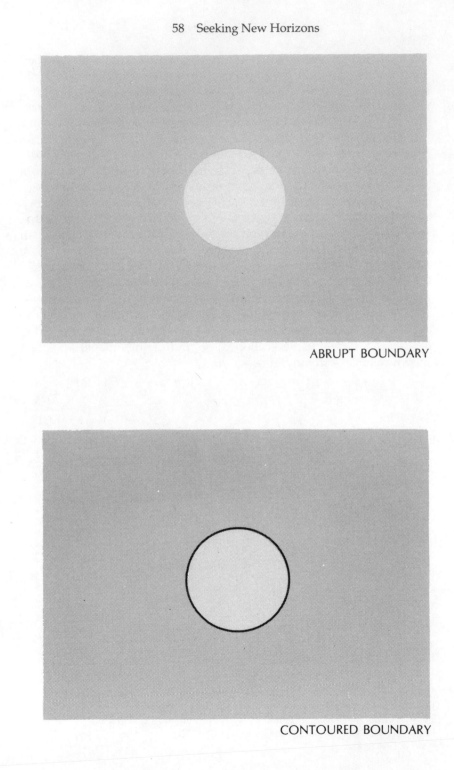

ABRUPT BOUNDARY

CONTOURED BOUNDARY

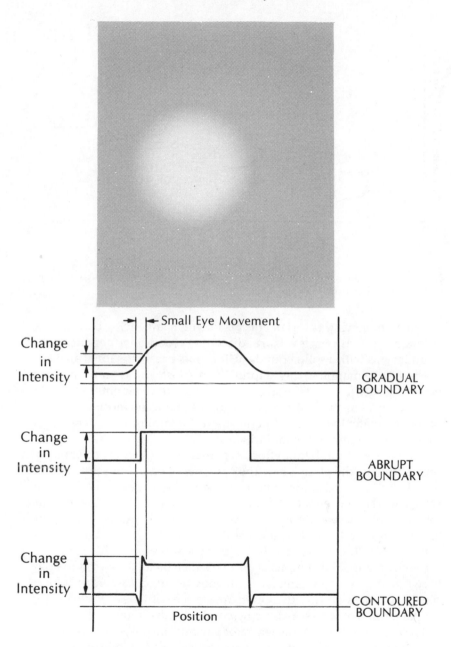

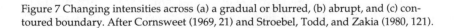

Figure 7 Changing intensities across (a) a gradual or blurred, (b) abrupt, and (c) contoured boundary. After Cornsweet (1969, 21) and Stroebel, Todd, and Zakia (1980, 121).

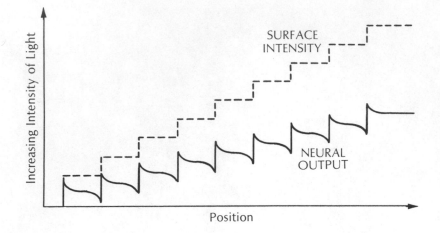

Figure 8 Surface intensity and neural output along a stepped surface, as across a gray scale. After Schiffman (1976,231,Fig.14.3bc).

In map making, this sort of effect is useful in enhancing boundaries, as long as the two tonal levels are adjacent. As soon as we demarcate the boundary, as with a solid boundary line, or separate the tonal areas, as in choropleth mapping, this helpful effect is lost, and we must depend upon other attributes of the areas to discriminate and identify them. As a result, the perceptual differences in area symbols on most maps must be much greater than theory might suggest in order to compensate for these diverse environmental influences.

The recurrent inhibition effect, together with eye movements, obscures from us the place where the optic nerve passes out through the rear wall of the eyeball, the blind spot[1] and the maze of blood vessels which overlay the retinal rods and cones. Without these two factors, an image would be seen through a network of blood vessels, as in figure 9. If we were to hold the eyeball still, and create what is known as a *stopped image*, the blood vessels could then be seen.

Obviously, it is the motions of the eye which break down the analogy between the eye and camera. With movement, a camera produces blurred pictures. Only with motion can we see with the eye because the retinal mosaic of receptors is being continuously "refreshed."

The eyes move in various ways for various purposes. The two most important movements for our purposes here are called tremors and saccades.

Tremors, or nystagmus, are the very rapid, small amplitude motions on which the very basis of vision rests. Typically there may be from 30

Figure 9 Network of blood vessels superimposed over a portrait; what the world should look like! (Cornsweet, 1969, 21) Permission granted by SRI International.

to 100 per second over amplitudes up to a minute of arc (Riggs, 1971, 386). Saccades are the slower, larger jumps which the eye makes in normal looking behavior, such as in inspecting a scene. Typically they are a quarter of a second in duration (Castner and Eastman, 1985, 33), and from a few degrees of arc to perhaps twenty degrees of arc when the head begins to move.

The motions of the eyes can be recorded and analysed in various ways and for various purposes. Cartographers have had a mild interest in them, a story which has been reported on elsewhere.[2] For our purposes, it is sufficient to provide a short summary of one aspect of that work.

In examining large numbers of eye movement records, particularly those taken in task-specific viewing situations, one is impressed with the great efficiency with which an individual can examine an image in search of a particular piece of information or in search of clues that will allow them to draw some kind of conclusion. By efficiency, I mean the

Figure 10 *The Unexpected Visitor* by Il'ya Yefimovich Repin. With thanks to the
Gosudarstvennaya Tret'yakovskaya galereya.

relatively small number of areas that are fixated given such a large
number of possibilities that an experimental graphic may provide.

The early eye movement research seemed to be intent upon studying
the places on which the eye fixated. But in task-specific situations, the
question is raised as to whether it is more revealing for the experi-
menter to consider what *was not* fixated upon, rather than to expend a
great deal of energy attempting to determine with great precision what
was. This suggests thinking of the visual system as a dual one. Foveal
vision involves the inspection of fine detail, while peripheral vision is
used to find targets that are likely to yield useful information when
closely inspected.

One of the most exciting sets of such records is that produced by
Yarbus (1967). Subjects were recorded looking at a painting in both an
unconstrained or free-viewing mode and in seven different task-spe-

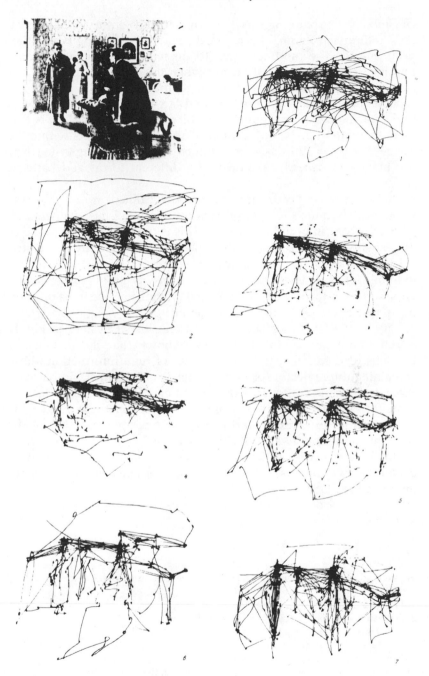

Figure 11 Eye movement records in free examinations (Yarbus, 1967, 172). Courtesy of the Plenum Publishing Corp.

cific modes. The painting was by the nineteenth century social realist, Il'ya Yefimovich Repin, of a political prisioner returning home after years in exile. The painting, figure 10, depicts the man being ushered into his living-room at that instant when the family's disbelief, that such a tramp could be allowed in, changes to recognition that the figure is someone very dear to them. By and large, the records of viewers under free examination (figure 11) exhibit many overall similarities that seem to relate to the strong structural and tonal contrasts within the image. But when subjects were assigned specific viewing tasks, their eye movement records take on vastly different forms and patterns (figure 12).

Obviously, in each case, the mind has "determined" in some way the information required to answer each question and has evaluated all *potential targets* in peripheral vision. This is the only assumption that can explain why large areas of the images were frequently not fixated upon. Those areas were being processed peripherally and found wanting in terms of targets likely to contribute to answering the experimental request. Hence the idea of a dual visual system; not one in its physiology *per se*, but in its cognitive functioning.

This sort of evidence suggests that the visual system is highly selective and purposeful. It has been shown that the majority of fixations tend to fall within image areas of high information, those containing unpredictable or unusual details. In contrast, simpler and more predictable contours, although as clearly recognizable as the more informative ones, are seldom fixated. Rather these predictable outlines and regular textures are processed in peripheral vision, by the parts of the retina away from the fovea. As a result, the relevant, informative areas of test stimuli can be located and fixated upon within a matter of two seconds or less from the start of a presentation (Castner and Eastman, 1984, 108).

The evidence of selective vision suggests two quasi-mutually exclusive alternatives:

Figure 12 Eye movement records under the following conditions or questions (Yarbus, 1967, 174):
1) Free examination.
2) Estimate the material circumstances of the family.
3) Give the ages of the people shown.
4) What was the family doing when he arrived?
5) Remember the clothes worn by the people.
6) Remember the position of the people and objects in the room.
7) Estimate how long the visitor has been away.
Courtesy of the Plenum Publishing Corp.

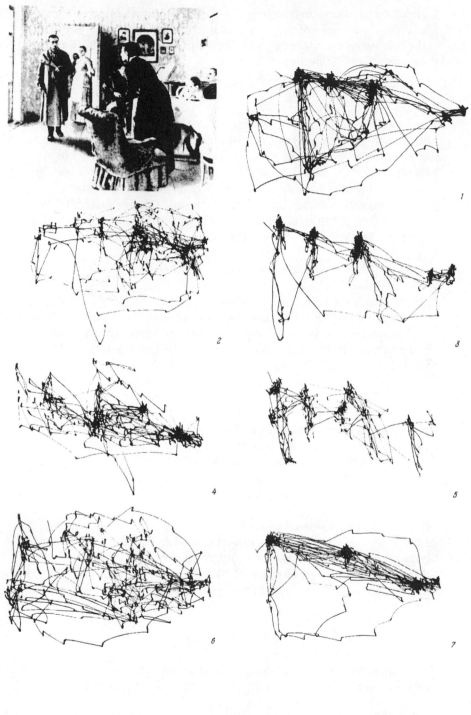

1 that the visual system is aware of its information needs either a) prior
   to viewing, because of memory, expectation, or experience, or b) after
   the initiation of viewing, because of the process of organizing stimulus details into coherent structures; or
2 that the visual system reacts in a deterministic way to the graphic
   properties of the stimulus.

These are, of course, questions of considerable theoretical importance.
But as with most things, the truth probably lies somewhere between
these alternatives. To understand the distinction between the two
alternatives, it is useful to consider more closely two contrasting
viewing modes: (1) "spontaneous looking," in which the viewer has no
predetermined information needs (as in figure 11 above); and (2) "task-
specific viewing," in which the reader begins his visual examination
with some specific goal in mind (as in figure 12 above).

SPONTANEOUS LOOKING

Spontaneous looking refers to the visual sampling strategy employed
by individuals when task constraints are absent — an experimental
situation frequently termed "free examination." Involuntary fixations
are attracted differentially across a map by three kinds of image
properties: physical, Gestalt, and cognitive.

Physical Properties

These refer to element dimensions such as size, position, isolation, or
color. Brightness contrasts, edges, and sharp angles (Kaufman and
Richards, 1969) also influence fixation activity.

Gestalt Properties

These refer to those visual properties that arise from the juxtaposition
of more than one symbol or stimulus element. These include various
Gestalt properties (such as proximity, similarity, and interparallelism)
which may arise from clusters of like symbols or from the natural
convergences or divergencies between adjacent contours of unlike map
symbols.

Cognitive Properties

These refer to strong intellectual connotations or associations that
symbols may have for the map viewer and which provoke fixation

Figure 13 Eye movement recording of a subject's free examination of Shishkin's painting *In the Forest*. From Yarbus (1967,183). Courtesy of the Plenum Publishing Corp.

attention. The associations may relate to pleasure or novelty in a particular context, even when attractive physical dimensions are absent. For example, in figure 13 the concentration and direction of fixations on the trees are probably influenced by the physical properties of their brightness contrast and their linearity; those on the human figure must be attributed to cognitive properties.

All this suggests that during free examination there is a variety of stimuli for eye movement activity and it may be fairly deterministic. However, more basic perceptual processes may be operating. One may be the simple epistemic need to understand the display one is viewing; here eye movements would reflect the viewer's cognitive exploration of the display. It has been noted (Kahneman, 1973, 78-9) that the stimulus characteristics which appear attractive to the visual system are also those which determine figure-ground relationships. Thus figural dominance, advancing colors, isolated or contour-rich stimuli are all attractive to the eye and help establish a figure separate from a ground. Neisser (1967,89) considers the breakdown of the visual display into its basic figure-ground structure as a very early part of the viewing sequence — a "pre-attentive," holistic operation which works on the gross characteristics of a visual display.

If such an early pre-attentive figure-ground breakdown does occur, we would expect subsequent fixation activity to work within that figural structure as part of the natural exploration of the image. However, the pattern of fixations as seen in an eye movement recording masks the important fact that the attributes of "attractiveness" or "informativeness" must have been detected previously in peripheral vision.

TASK-SPECIFIC VIEWING

The second viewing mode, task-specific viewing, refers to the visual sampling activity undertaken by the eyes during visual problem solving. The recordings in figure 12 reflect the changing need for information and thus the "attractiveness" of particular symbols, which relates not to their physical or Gestalt properties *per se* but to how they pertain to the specific visual task at hand. DeLucia (1974, 1976) reported similar findings by noting that as the task facing the viewer became increasingly specific: (1) the complexity, the apparent "randomness," and the spatial extent of the scan path decreased and (2) graphic influences decreased.

These findings support the notion of an epistemic or task-related override function in the acquisition of visual information. As the task facing the viewer, or his need (or motivation) to gain a particular piece

Figure 14 A pair of complex symbols: one is composed of a single line turning upon itself; the other is composed of two interwoven lines. After symbols devised by Marvin Minsky and Seymour Papert and reported on by Julesz (1975, 34-5).

Figure 15 A pair of simple figures whose differences are immediately apparent. Comparable to symbols designed by Julesz (1975, 35).

of information becomes more specific, the prevalence of this override function is strengthened. In this situation, the role of peripheral vision must be even more important than in free examination, for not only are possible fixation sites located, but also their significance for possible scrutiny is established in terms of the viewing goal at hand. In the end, therefore, it may well be the viewer's information needs at the time of viewing which will determine the allocation of attention (and eye fixations) to any particular part of the display.

PURE AND COGNITIVE PERCEPTION

The identification of two ways of processing a visual image, through foveal and peripheral vision, suggests another way of differentiating these visual processes.

In figure 14 we see a pair of complex symbols: one is made up of a single line turning upon itself; the other is made up of two lines interwoven together. It has been shown that subjects are visually unable to detect this distinction without attempting to trace the lines in order to ascertain their connectedness (Julesz, 1975). On the other hand, subjects experience no difficulty in immediately seeing that one of the symbols in figure 15 is made up of two separate lines.

The detection of the differences in the two complex figures is accomplished only by a conscious, deliberate, cognitive attempt to solve the specific question of whether or not there are two lines. In contrast, the detection of the difference in the pair of simple figures involves no *conscious* effort — it is immediately evident and effortless. The mind is involved, as it is in all aspects of vision, but the information was processed so quickly and a decision was reached so effortlessly that the reader may not have been aware of the processing going on. Julesz calls this a *perceptual* process — one that is performed spontaneously without help from cognitive processing that involves *scrutiny*. Later, he defines it as "effortless or immediate perception" (1981). Gazzaniga (1985, 118) refers to a similar term, "unconscious processing," used in experimental psychology. Julesz also points out that there is in essence a continuum of figures between the complex and the simple pairs; as the figures become more ambiguous, there is a limitation to the perception of this connectedness. As a result, the gradual overloading of the perceptual system requires that some higher form of cognitive processing must be used. Hence the distinction between pure and cognitive perception.

The idea of a tiered visual system can be found in other places and stated in other terms. Ehrenzweig (1969) devotes an entire chapter to what he calls "unconscious scanning." He attributes this low level vision to the ability to grasp "in a single undivided act of comprehension" data that would be "incompatible" with conscious perception. In other words, unconscious scanning and conscious perception are mutually exclusive processes. The latter "enforces" the selection of a definite Gestalt as a figure. Unconscious scanning of the total visual field, on the other hand, can detect "far-flung structures," thus offering a great number of choices for any creative search. He also presents evidence that unconscious scanning operates subliminally and far faster than conscious scrutiny and registers details irrespective of whether they belong to the figure or to the ground. Unconscious scanning pays more attention to textural and background elements.

Timothy Gallwey (1974) also talks about focusing one's conscious energy as it applies to learning to play the game of tennis. Since this game has an important visual component, it is easy to see how his thoughts relate to our discussion. First, he hypothesizes two "personalities" existing within the many tennis players who talk to themselves: the one doing the talking and giving the instructions, "Self 1", and the other, the silent, nonverbal one who seems to be performing the action, "Self 2." He points out how well Self 2 alone can perform the thousands of operations that are involved in hitting a tennis ball. For example, in returning a serve:

In order to anticipate how and where to move the feet and whether to take the racket back on the forehand or backhand side, the brain must calculate within a fraction of a second the moment the ball leaves the server's racket approximately where it is going to land and where the racket will intercept it. Into this calculation must be computed the initial velocity of the ball, combined with an input for the progressive decrease in velocity and the effect of wind and of spin, to say nothing of the complicated trajectories involved. Then, each of these factors must be recalculated after the bounce of the ball to anticipate the point where contact will be made by the racket. Simultaneously, muscle orders must be given — not just once, but constantly refined on updated information. Finally, the muscles have to respond in cooperation with one another; a movement of feet occurs, the racket is taken back at a certain speed and height, and the face of the racket is kept at a constant angle as the racket and body move forward in balance. Contact is made at a precise point according to whether the order was given to hit down the line or cross-court — an order not given until after a split-second analysis of the movement and balance of the opponent on the other side of the net. (Gallwey, 1974, 49-50).

With someone like Pancho Gonzalez serving, you have approximately .613 seconds to accomplish all this! Obviously Self 2 must be quite adept. Seiderman and Schneider (1983,17-18) make a similar observation about hitting a baseball – "the single most difficult feat in sports." If a major league pitcher throws the ball at eighty miles an hour, it will take 4/10 of a second for the ball to reach home plate. Since it takes 2/10 of a second just to swing the bat, less than that amount of time is available in which the batter must decide whether or not to swing at the pitch. Only batters with extremely good dynamic visual acuity can detect the spin on the ball as it leaves the pitcher's hand – revealing information that will help make the decision to swing and how to swing.

However, you do not have to be a tennis or baseball player to observe your tiered visual system in action. Just walk with someone from one room in a building to one in another building, through doors, down stairs, around obstacles, all the time talking to and maintaining eye contact with that person. The secret to successfully playing a sport like tennis lies in allowing Self 2 to take control of the basic strokes and manoeuvre about the court. Self 1 can then tend to other matters such as tactics, watching the clock, or whatever.

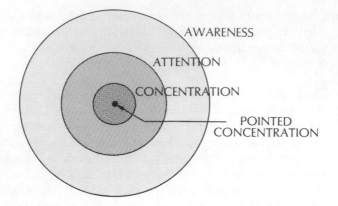

Figure 16 Four levels or areas of visual attention. After Gallwey (1974,98).

It is interesting that Ehrenzweig (1969,35) makes a similar distinction in relation to creative viewing. He states a general law that we can only gain significant insights when we allow the unconscious scanning process to take over from the conscious one. Dee Coulter (1986) gives an example of this process when she describes herself reading some non-narrative material such as the titles of all the paperback books in print. She actually did this one day during her doctoral studies in order to relax and clear her mind for some more important work to come. What surprised her was that several days later she became aware of the different topics that were *not* being written about! This would be an impossible conscious task: "How do you go about," she asked, "looking in a book for information that is not in there?" But left alone, the unconscious mind can come forth with ideas which were never consciously sought.

Timothy Gallwey also describes four concentric areas of visual concentration used in playing tennis (figure 16). There is certain information which must be picked up in a very restricted area of view, for example crucial information about the position of the ball and its rotation. Slightly less crucial is information about the position of the ball relative to the court and net; this comes from a slightly larger visual field. Then there is information about the movements of the opposing players; attention to this may take place over a still larger area. Finally, one may have an awareness of spectators or noise from an adjacent court or a plane overhead. Of these areas, the first requires close scrutiny; the other three are monitored in peripheral vision.

While it may not be possible to correlate perfectly the peripheral/foveal vision distinction with the perceptual/cognitive distinction of Julesz, it is a useful association to make. Indeed, the literature reveals other pairs of processes which describe the same kind of image break-down, for example, the distinction between what Julesz calls pure and cognitive perception (Julesz, 1975). These include the terms "filtering and matching" (Beller, 1972); "syncretic and analytic" (Ehrenzweig, 1969); "soft eyes and hard eyes" (Leonard, 1977); "pre-attentive and focal attentive" processing (Neisser, 1967); and simply "Self 2 and Self 1" (Gallwey, 1974). This same breakdown is also reported by Gazzaniga (1985, 99) who notes that there are other systems coincident with language that do the serious computing and decision making — language merely reports on these other processes.

These pairs can also be generally related to the concepts of "pattern perception and figure perception" (Castner, 1979a, 156). The former process in each pair is largely associated with and can be accomplished by peripheral vision; the latter with central or foveal vision and is the focus of conscious attention. They are, of course, not separate processes for they work hand in hand.

THE NATURE OF PERCEPTION

If it is change in the visual array which excites the eye, then obviously it is the patterns of change which are the attributes of an image or scene that convey information about it. Our interpretations of the "meaning" of the image or scene are derived from that information.

James Gibson is one psychologist who has been particularly interested in how these interpretations are made. The crux of the problem for him is the observation that constancy is the tendency to perceive an object as unchanging despite the changing sense impressions that we may have of it (Gibson, 1965, 60). What follows is a brief reconstruction of Gibson's hypothesis in this regard.

The three most studied constancies in vision are those of size, shape, and color. Size constancy refers to the fact that we can judge the size of an object fairly well at a variety of visual distances. As a result, we can compare the size of objects nearby with those far off. In addition, the space between things is also a constant in experience. Shape constancy refers to the fact that we see the shape of a facet of an object correctly even when it is tilted or inclined to our line of sight (when its image is foreshortened). Color constancy refers to the fact that the perception of surface color or value does not seem to vary much despite large changes in the absolute illumination of the surface, and hence the sensation of that surface.

As a result of these three constancies, we are aware of the surface of our environment and the "layout" of these surfaces (Gibson,1965,61). However, we are not so much conscious of the patchwork of colors, edges, and textures (as determined by the laws of perspective) as we are of perceiving the environing surfaces with their edges, corners, slants, convexities, concavities, and interspaces. These are, of course, the pathways and obstacles, the places and things, the goals and the dangers of the terrestrial world. They are identified by their surface properties, including texture and differential reflectance of light. They have to be seen as constant, where they are, in order to be identified for what they are." (Gibson, 1965, 61). Thus he declares the central problem of perception has to be how we see depth and distance, the so-called third dimension of space.

There are various techniques in art and descriptive geometry which assist us in providing in a drawing the necessary cues and clues to this third dimension. But these techniques do not explain in themselves why they allow us to perceive the constant layout of space. In other words, whenever an object moves relative to an observer (or an observer moves from place to place), the pattern of the observer's field of view undergoes a perspective transformation. Gibson calls this pattern the optic array which determines the observer's retinal image. He goes on to note that unless we hold our heads unnaturally still and fix our eyes, our visual field is alive with change or "transformation." A sensation of "form," then, is an extremely rare event in life, for the eye is continually being stimulated by transformations. In addition, sense impressions are highly unstable and interrupted by the blinks and saccadic movements of the eyes and by the narrowness of the foveal field. Thus, he concludes, the stability and continuity of perception cannot be based on sensation. If it were, it would be more related to the cone of visual rays approaching the eyes than to the "panorama which surrounds us."

Historically, there have been many explanations for the stability and continuity of perception. Gibson classifies them roughly into three types which depend upon (1) intuition, (2) experience, or (3) some kind of sensory organization. The explanations all assume "the poverty of the senses" and look for some special process in the mind which supplements them (Gibson, 1965, 64).

Intuitive explanations assume some sort of inborn or nativist capacity of the human brain, a preconception, for example, that certain objects are rigid, or that space is constant and Euclidean in nature. Experiential explanations assume that visual constancy is learned, so that memory or experience supplements or interprets the sensory data. This explanation has the appeal that it may be possible for this interpretive process to be improved with the "accumulation of memories." The third kind of explanations, those dealing with sensory organization, arise from the critical question addressed to the second: How could a child learn to see an object without ever having seen an example of one? Such a theory of association, or of learning cues, cannot explain the object unless the essence of the object has been presented at some previous time. The question was raised by Gestalt psychologists who proposed that a process of sensory organization, instead of association, accounted for the constancy of perception. The nature of the nervous system is such that organization takes place "spontaneously," so that a visual field is never wholly meaningless, even at the start (Gibson,1965,65).

Among the most explicit organizations arising out of vision are the various so-called laws of Gestalt Forms. These include the tendency to see those things as wholes which have a) elements which are near to each other; b) elements similar to each other; (c) continuity of contour; (d) uniformity of direction; (e) a tendency to enclose areas; and (f) the unity suggested by experience. These have come to be known respectively as the law or factor of (a) proximity; (b) similarity; (c) good contour or common destiny; (d) common movement or interparallelism; (e) closure; and (f) experience (Hamlyn, 1969, 45).

Koffka (1935, 91f, 222f) noted that we do not see separately a retinal size and then a distance, but rather we see all at once a "size-at-a-distance." The relationship between size and distance is an invariant relationship in experience — as one changes, the other changes proportionally. Similarly, shape is reciprocal to slant or positional attitude. Gibson then takes the next step by stating that size-at-a-distance and shape-at-a-slant may actually be given as invariants within optical stimulation. He further suggests that the proposed Gestalt laws of organization have not been verified experimentally (even though some kind of organization in perception is a fact) because we are looking for explanations internally in the mind; perhaps, he asks, they are to be found externally?

Perceptual constancy then, in Gibson's terms, is the invariance of perception even with varying sensations. In other words, the environment is constant or stable, stimuli from it are ever-changing, and thus our sensations of these stimuli are also changing; but our perception of the environment is constant! For example, the third dimension of space can be thought to be "lost" in a two-dimensional visual image, but through some activity of the mind it is restored in visual experience. One might say then that the various techniques of art and descriptive geometry provide us with ways to "embed" or "bury" clues and cues into two-dimensional images so that the three-dimensional experience can be "retrieved" by the perceiver.

It might require a great leap in logic to assume that the process of perspective drawing is one that creates the invariant attributes of a three-dimensional space. But certainly it must be of great value to be in command of any procedure which allows one to knowingly create a distorted image which then allows the constancy or stability of a space to be experienced again.

Gibson goes on to point out that one perspective view of an object carries some information about the object. A second view carries even more. The total experience of that object would be bound up in a whole series of views which vary systematically and reversibly. Thus the image of the object on the retina can undergo a continuous sequential perspective transformation even though the object itself remains unchanged. This he called the invariance of perception.

In the past, it was assumed that these ever-changing forms of the stimulus are "a chaos which cannot possibly contain the solid form of the object." (Gibson, 1965, 67). Gibson suggests that they can, namely, that the true shape of the object may be implicit in the orderly series of transformations. If this is so, then the order or structure of the object does not have to be imposed on any single array of light — it is already there in the sequence of images, the "sequential stimulus."

On the basis of these arguments, Gibson proposed a new approach to the invariance of perception which suggests that sensory experience and perceptual experience are two different things. The former he calls a special self-conscious kind of awareness while the latter is unself-conscious and direct. The latter is not dependent upon the former. Perception is not mediated by sensations, nor based on sensations. This means that the stimuli causing sensations are different from those causing perceptions. The hypothesis is that the "flowing array of light" from an object or scene has two components, one of change and one of nonchange. Those of change relate to sensation; those of nonchange relate to perception. In 1966, he put it this way:

Continuous optical transformations can yield two kinds of perception at the same time, one of change and one of non-change. The perspective transformation of a rectangle, for example, was always perceived as both something rotating and something rectangular. This suggests that the transformation, as such, is one kind of stimulus information, for motion, and that the invariants under transformation are another kind of stimulus information, for the constant properties of the object. (1973, 43).

Object perception, then, does not depend on form-perception but on invariant detection. In other words, the research on the discrimination of forms and patterns has no relevance to the perception of the *environment*, only to *pictures of the environment* (Gibson, 1973, 43).

At this point, Gibson borrowed two notions from mathematics:

transformations and *invariants under transformation*. The former, in mathematics, is not simply change but the permanence of change. If we consider that stimuli are not static forms but "serial transformations," then we can regard the invariant component in a transformation to carry information about an object, while the variant component carries other information, such as that concerning the spatial relationship of the observer to the object. Thus the permanent properties of our environment — its texture, edges, layout, solidity, stability, and the fact of gravity — are undoubtedly specified by invariant properties in the "visual and tactual stimulus flux" (Gibson, 1965, 68).

Gibson also considered the question of whether the perception of an object can be aroused by a picture of the object. Since a picture presents no transformations, being still, it can display no invariants under transformation. This led Gibson to ask whether there could be any *formless* invariants in a picture. Could we learn to detect invariants in a frozen array as well as in a changing array? Is it possible that the information in a picture does not consist of the forms in that picture? Gibson came up with seven answers to these questions; four of the more interesting for our discussion are given here (Gibson, 1973, 44).

1 Young children, like primitive men, are probably not aware of forms as such until they learn to draw and to perceive by means of drawing. On the other hand, Gibson wonders if it is not the formless invariants of scenes or objects that a child is noticing and is attempting to replicate in his first scribblings of childhood (Gibson, 1978, 230). Thus in any picture of a cat, the young child sees *catness* and not the side view or front view of a cat, as the case may be. The front and side views specify the same cat and are seen as such.

2 In observing a caricature, like a political cartoon, we recognize the person depicted even though the image is a poor projection of the face and the detail is sparse. In other words, we do not notice the lines as such, but only the information they convey. The essential, invariant information about the person is preserved, and the realism of the image is not based upon an analytical match (Ehrenzweig, 1969, 8).

3 People who put on special spectacles, such as prisms, that distort, rotate, or invert the retinal image are able to adjust to their distortions and learn to operate successfully in their new "visual" environments. Thus, their *perception* of objects and of the environment tends to become usable over time. This suggests that information about the environment is not carried by the form of the retinal image, and that perception is based on something else.

4 In reading, it seems clear that we do not notice or even attend to the peculiarities of the graphic elements of the letters on the page, but

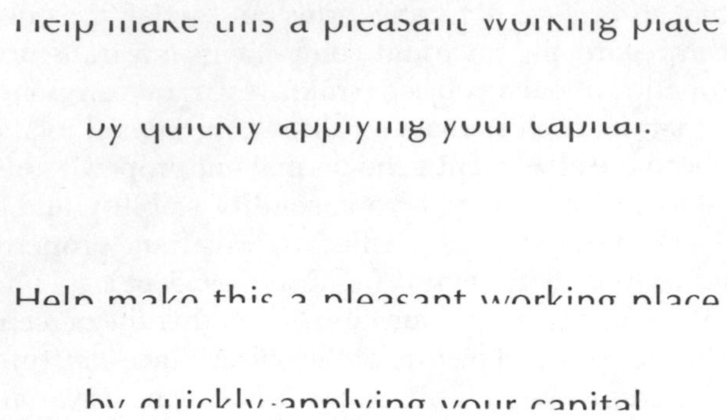

ι ιeιp ιιιακe ιιιɔ a pιeaɔaιιι wυικιιιg pιace

υy yuιcκιy appιyιιιg yυuι capιιaι.

Help make this a pleasant working place

by quickly applying your capital

Figure 17 Sentences in which the top half a) and the bottom half b) of the letters have been obliterated. The loss of the top halves appears to be the more damaging to comprehension.

only to the distinctive combinations of letters that form words (E.Gibson, 1969, Chap. 19). Evidence of this can be seen in sentences where the upper or lower halves of the letters have been obliterated, as in figure 17, and in our ability to read a great variety of handwriting.

Gibson is certainly vague about the precise nature of the formless invariants. He provides some examples[3] but also notes that they cannot be put into words or symbols. From what he has said, we should be looking for pairs or continua of properties, such as the aforementioned size-at-a-distance. Figure 18 lists some possibilities that may convey invariant information. Another psychologist (Garner, 1981,127), in his discussion of the analysis of unanalysed perceptions, speaks of the configural properties of symmetry, repetition, redundancy, intersection, conjoining, and angular separation. Muller (1976), a cartographer, has also described some collective visual stimulus dimensions to which he believes map readers are responsive. These include blackness, redundancy, aggregation, compactness, complexity, and contrast.

CONCLUSION

Obviously the stimulus properties noted above are essentially those of topological space and of the Gestalt laws of visual organization. If we

Straightness/Curvature	Continuity/Discontinuity
Termination/Continuation	Intersection/Connection
Parallelism/Convergence (Divergence)	
Closure/Openness	Alignment/Offset
Similarity/Dissimilarity	Proximity/Isolation
Occlusion/Separation	Coincidence/Occlusion
Symmetry/Assymmetry	Perpendicularity/Obliquity

Figure 18 Pairs or continua of properties of graphic elements that may convey invariant information.

are to believe Piaget, they are properties which children are sensitive to long before they can deal with the Euclidean properties of space. They are also generally invariant in the transformations we think of in projective and Euclidean space. They seem also to mesh with the relational factors in experience which Suzanne Langer regards as either intuitively recognized or not at all. These include such formal charac- teristics as distinctness, similarity, congruence, and relevance; they are "protological in that they must be seen to be appreciated" (1957,166).

Whatever the case may be, these kinds of properties provide a more specific language for use in searching for structures and relationships within graphic images. As well, skills in both foveal and peripheral vision would seem to enhance our powers to observe them (our abilities to analyse graphic images and our environmental surround- ings). Whether or not we label them as formless invariants is unclear and perhaps unnecessary. Gibson readily admits to the shortcomings of his hypothesis and has over the years abandoned some of its aspects[4]. It is clear that the psychology community has not accepted uncritically Gibson's radical hypothesis, although in recent years there are indica- tions of renewed interest in the implications of his ideas. However, the idea of invariant dimensions is an intriguing one. It supports the notion that there are all sorts of structures in complex graphic images (besides the obvious ones which we have identified heretofore) that may de- scribe more of the implicit relationships within those images. This suggests that some sort of systematic visual analysis can be used to reveal those structures in images.

CHAPTER FOUR

The Music Connection

After all, a child's first response is to sound. In fact, his first response *is* sound. His next response is to rhythm – patting, rubbing, rocking, lullabies, etc. All mothers know this. How did educators forget so soon? And why isn't the use of playing, singing, and rhythm and movement a part of the core curriculum?

Margaret Fish (1977)

In short, although the notes of a melody are its elements, they are not the basic units in terms of which melodies are perceived. Rather, it is the relations among the notes that are the essential, irreducible properties of a melody.

Anne D. Pick (1979, 147)

If musical ability is non-verbal, seated in the right half of the brain, we must approach it on a non-verbal level.

Isabel Carley (1977c, 84)

After seeing, listening is perhaps the richest source of environmental information for human beings. The loss of either system would obviously necessitate extensive physical and mental adjustments. However, the concern here is not to contrast the nature of those adjustments, but to consider the many interesting parallels between these two perceptual systems – seeing and listening – and, by extension, between cartography and music.

First of all, we can assume that listening with precision and discrimination is as basic a life skill as is seeing with precision and discrimination. Needless to say, they are both as fundamental in this modern world as skills in the use of language and mathematics. Thus every school curriculum should provide for the development of general abilities, and some specific skills, in all four of these areas.

It must also be recognized that many fundamental experiences in spatial awareness, direction, and orientation are given by any music education method that involves the body and mind in movement and dance. In particular, those developed by Suzuki, Orff, and Jaques-Dalcroze suggest ways in which abilities and skills in visual perception and spatial awareness might be taught in the areas of cartography and geography. The basis of this claim rests on the close resemblance between improvization in music and some of the thought processes in geography; processes that go into thinking about spatial relationships and their representation on maps. By improvization, here, is meant the manipulation of specific perceptual variables within some well-defined limits. The geographic thought processes can be defined collectively by the term "mapping": They include the cognitive activities of (1) thinking about some aspect of the world; (2) seeking the significant dimensions or characteristics which describe that aspect; (3) establishing a communication goal(s); (4) considering the various modes and forms of representing those significant characteristics; and (5) executing a graphic representation.

This chapter attempts to clarify an unexpected relationship among music, cartography, and geography and also tries to suggest a number of a) teaching activities that make use of this relationship for improving skills in visual discrimination and spatial awareness and b) place associations that can be found in much music, particularly that of the folk and classical traditions.

SEEING AND LISTENING

Let us examine some interesting parallels between seeing and listening. In both, the more skilled we become as perceivers, the better able we are to identify an increasing number of properties that remain invariant even as they are presented in increasingly complex and interesting transformations (Pick, 1979, 163). By analyzing the structure of a stimulus, and manipulating it in controlled ways, we can discover the essential properties of that stimulus. Over time, the crucial information concerning the invariant properties of an event can be known (Pick, 1979, 146). It is also possible to make in the auditory mode the same kinds of distinctions that I made in chapter 3 concerning the processes of perception and cognition, or peripheral and foveal vision. Let me explain this with an example very close to home.

My wife and I enjoy listening to classical music, but, in part, we respond to quite different aspects of it. As a trained musician and music educator, she is often able to discern, follow, and reproduce (both musically and verbally) many of the progressions of the melodic and harmonic lines. These can become quite complex: for instance, there are inversions (in which the rising and falling lines of a melody are reversed); retrogrades (in which the sequence of intervals is reversed, as from beginning to end); and retrograde inversions (where the pattern of intervals is reversed both in direction as well as in sequence) (Pick, 1979, 161). The ability to detect these kinds of relationships with all the other things which may be going on in a full-scale symphonic orchestration requires a listener to have the auditory equivalent of excellent foveal vision, that is to say the ability to focus on extremely fine "detail" – what Ehrenzweig (1969, 32) calls vertical attention.

In contrast, I can follow major melodic lines, but, by and large, I have great difficulty in reproducing them, vocally or describing them in words. On the other hand, without any formal musical training, I am reasonably adept at identifying the composer of an orchestral piece. I do not accomplish this by recognizing a particular melodic line, although at times this helps to confirm an identification. Rather I respond holistically (or horizontally, as Ehrenzweig would say) to the overall texture of the orchestration. Dowling and Harwood (1986, 160) would describe this response as an undifferentiated, unconscious perception of the invariants or structural constancies underlying surface change in local pattern features – regularities of temporal organization such as of beat, tonal scale, instrumentation, or note spacing in pitch and time. For me, the name of the composer is particularly clear in the mixture of sounds produced by an orchestration of the composer's music. For example, a Rachmaninoff symphony has a unique sound, one that is easily discriminated from that of another Russian composer, Tchaikovsky, who wrote a little earlier but in the same cultural setting, or from Sibelius, who composed at about the same time but in Finland, in a different culture, although one in close geographic proximity. Thus I consider that my aural equivalent to peripheral vision is quite acceptable.

There is no question that horizontal and vertical attention are two valuable listening skills (vertical attention to the progression of the notes themselves and horizontal attention to the interaction or textures among the coincident notes). How these skills are acquired in a formal educational setting is a moot question. For, after all, progress in their development cannot be measured by how many songs one might have learned to sing or how well they might have been sung. Rather, the acquisition of these skills seems to relate to another level of cognitive awareness or understanding. Two teaching philosophies that have become well known in the last decade or two in North America may offer insights into this awareness. These are the processes initiated by Shinichi Suzuki and Carl Orff.

SUZUKI AND ORFF

Shinichi Suzuki and Carl Orff, educators from two quite different cultures, have developed methods for teaching music to young children that utilize the same perceptual skills children have used in mastering their own language (their mother tongue) – critical listening, repetition, and (particularly with Orff) improvization. There is a great deal of evidence of the success of these two approaches to music education that essentially develop skills in listening, aural perception and discrimination, and foster an understanding and appreciation of the art of music. Would not such skills have a place in the visual arts? Perhaps improvization in music is a closer analogue to considerations that are taken in map design than it is to some of the techniques we currently employ. This is suggested by the fact that in music composition and map design there is, in communicating ideas, a need for the controlled or disciplined use of the various basic design components, be they musical notes or graphic elements.

By discovering, and applying a cartographic analogue to improvization in music to geographic teaching, perhaps students will more easily identify the basic perceptual concepts which underlie successful map reading and spatial visualization. In other words, by working with these concepts within the more disciplined constraints of map design and cartographic communication, similar levels of success can be achieved with young children in developing their abilities in visual perception and spatial awareness. By improving these general abilities, more meaningful long-term results may be gained among our future adult map users than through efforts at remedial training of adults in various specific skills of map use and spatial mobility.

Shinichi Suzuki

Dr Suzuki is a master violin teacher and educational philosopher.[1] Born into the family of a violin manufacturer in 1898, he began playing the violin as a young adult. Eight years of study in Berlin in the 1920s led to a professional teaching career in Japan. His astounding success with very young children led to the growth of a movement known throughout the world as Talent Education. It has seen particularly strong growth in post-World War II Japan. It spread to North America in the sixties, and to Europe in the seventies.

His method sprang from his observation: "Oh – why, Japanese children can all speak Japanese! The thought suddenly struck me with amazement. In fact, all children throughout the world speak their native tongues with the utmost fluency. Does this not show a startling talent?" (Suzuki, 1969, 9). Suzuki had had a difficult experience with learning a language when he studied in Germany. There, of course, it was the German language with which the young children were fluent. Dr Suzuki believed their success with language demonstrated a remarkable ability fostered by the fact that the children were surrounded by language sounds from birth. He reasoned that if children were surrounded by musical sounds to the same degree, they would develop an equally remarkable ability in music. His method, then, applies this idea to the teaching of the violin; it has since been extended to the other string instruments, the piano, and flute.

Two principles are regarded as fundamental to the method: (1) the child must be helped to develop an ear for music; and (2) from the very beginning, every step must, by all means, be precisely identified and thoroughly mastered.

The first of these principles is fostered by a variety of teaching techniques. They include the creation of as favourable a teaching environment as possible, a great deal of positive reinforcement of steps achieved, the building up of a repertoire of progressively related pieces in performance, and an environment where a great deal of time is devoted simply to listening. Suzuki recognized that not every child learns at the same rate. Thus "we must give the stimulus that will enable him to learn at his own rate" (Suzuki, 1973, 109) and provide a teaching environment in which (1) the child can discover that rate; and (2) the teacher will be comfortable in allowing the child to work (at that rate). Too often we may equate "good" with completing a piece of work, not in doing it well (Suzuki, 1973, 113).

It is perhaps significant to mention that note reading, rhythmic reading, and such are not introduced initially in the program but only after the student has had an opportunity to *experience* the attributes or dimensions of the music which is being studied.

Carl Orff

Carl Orff, born in München, Germany in 1895, is perhaps better known as a composer of music, particularly for the theatre (Liess,1966), and of the secular cantata *Carmina Burana*. While his career differed markedly from that of Suzuki's, there are some startling parallels in how they viewed the learning of music, broadly speaking. For our interest, perhaps the most significant event in Orff's career was his meeting and collaboration with Dorothee Günther in founding the Günther Schule in 1924. The school aimed at unifying the disciplines of dance and gymnastics. In speaking of her aims, Günther states:

I wanted to discover a method of reviving the natural unity of music and movement – music and dance; a method which would be available not only to a few natural artists but would solve the educational problem of awakening in everyone the sense of rhythmic movement, and of stimulating a love of dancing and music making – a general freedom of expression and receptivity. This unity of music and movement was not to be based on incidental and subjective experience but on their elemental relationship, in that they arise from a single source. (Liess, 1966, 17)

Clearly, as with Suzuki, the teaching of music is seen in a far broader realm than merely as skill training. Liess (1966, 61) goes on to describe that "this education to music ... inevitably extends to general culture, and thus becomes education through music." "Orff's teaching method makes it quite clear that his return to primal origins is no historical exercise, but an expression of living experience."

In much the same way that Suzuki has great respect for the child as a learner, Orff's approach recognized" that the child's world had its own perfection. The new task was to teach the child to develop as a personality in its own right, by working and learning of its own accord" (Liess, 1966, 56).

To accomplish this, Orff, with the aid of Curt Sachs and Carl Maendler, the great harpsichord and piano builder of the time, created a new range of barred percussion instruments – xylophones, glockenspiels, metallophones; to these they added unpitched percussion and record-ers. With these, and the voice – the first instrument (Wuytack, 1977) – the child is able to experience directly and thus discover the basic prin-ciples of musical expression. For this experience, the pentatonic scale – the most fundamental and innately human musical scale – is initially used. It is interesting to note that playing on only the black keys of a piano also provides experiences with a pentatonic scale, and with using a much simpler keyboard (Kodaly, 1957; *Fifty*, 1964). Orff's work with Günther, however, was initially intended for professional dancers and musicians, not for children. It was Orff's collaborations with Gunild Keetman that led to the adaptation of Orff's ideas and techniques both to these new instruments and to school use in what was to become known as the *Schulwerk* (Choksy et al., 1986, 93-4).

As with Suzuki, the graphic expressions of music in *Schulwerk* (the bar lines, signature, and notes) are not introduced until after the child has experienced the music, and not as a hurdle to be passed before music can be produced. For, after all, it is claimed that individual singing combined with listening to music can develop the ear's sensi-tivity to what is heard to such a degree that it is as though one were looking at a score (Choksy, 1981, 7). Even then, the elements of the

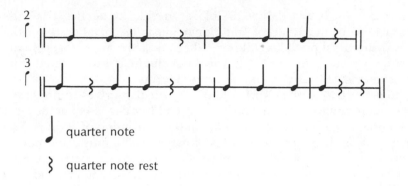

> $\Large{\text{♩}}$ quarter note
>
> $\Large{\text{𝄽}}$ quarter note rest

Figure 19 A line of quarter notes and rests for beginning recorder students. After Carley (1980,1).

musical line can be introduced gradually so that they can be utilized and understood without the presence of unexplained (and thus super-fluous) components which might at first be distracting (figure 19).

Music, in its origins, was a unity of sound, rhythm, movement, and word; these only become differentiated through experience. In the more general sense, then, Orff-Schulwerk is a successful and fully realized pedagogic method. It is a general procedure for guiding children through several phases of musical development by means of exploration, imitation, improvization, and creation (Shamrock, 1986).

By exploration is meant the discovery of the possibilities available in both sound and movement. By imitation is meant the development of basic skills in rhythmic speech and body percussion, in rhythmic and free movement through space, in singing, and in playing three kinds of instruments — unpitched percussion, the special Orff pitched percus-sion, and recorders as melody instruments. Body percussion involves such things as clapping, finger snapping, thigh slapping or *patschen*, foot stamping, and so on. By improvization is meant extending the skills with these components to the point where each individual can initiate new patterns and combinations as well as contribute to group activity based on this ability. By creation is meant the recognition and use of material from the previous phases into original small forms (two-part form, rondo, theme and variations, for example, and transforming literary and graphic material into some combination of natural or rhythmic speech, movement, singing, and playing instruments. One can easily envision here, as well, graphic displays.

Thus Orff-Schulwerk is a process- rather than a product-oriented methodology.[2] As such, the method does not produce the kinds of artifacts of success and effectiveness that we expect from school choirs and marching bands. Rather, its "place" in educational theory (and its attraction to thousands of traditional music teachers) rests on the opportunities it provides children to become active participants in the learning process. As a result, music is not taught, it is discovered in its most elemental, natural, and perceptual forms.

In essence then, the Orff method in particular strives to provide a classroom environment in which young children can discover the most fundamental or primitive elements in the perception of sounds. The emphasis is clearly on the process of experiencing music and not on performance or polished technical skill, although much of this is evident. By addressing the nonspecific development of children's abilities to listen with precision and discrimination, the method provides the basis for training, understanding, and the appreciation not only of music but also of language. Seen in this way, the Orff method appears to be making a contribution to music education that is far beyond that which we in North America, at least, tend to associate with music in the schools.

Assuming that a fundamental goal of our educational systems is the development of our most basic perceptual skills – listening and seeing – then perhaps there are similar concepts and methods in the graphic arts which can be directed towards developing perceptual skills in vision. In chapter 5, this question will be raised in the context of art education. In the meantime, it is sufficient to suggest that music education seems to bear a much closer and perhaps more significant relationship to education in cartography because of the restricted number of options available to both composer and map designer for communicating ideas. It is not coincidental that both Suzuki and Orff have chosen circumstances with limited options in which to allow students to work out and experience the basic contrasts and relationships among those options. Progress is then made by gradually increasing the number of options, and thus the number and type of contrasts and relationships which the child can create and thus experience. This is the essence of improvization.

THE NATURE OF MAPPING

It is now necessary to ask what parallel method might be employed in making or using maps. In this regard it is interesting to note how Robinson and Petchenik (1976, 4) elucidate the noncartographic meaning of "using a map" when they note that " it is possible to take isolated incidents, experiences, and so on, and arrange them intellectually so that there is some coherence, some total relation, instead of individual isolation." Later on (p.55), they remark that " 'arranging' seems to be a fundamental human activity, and man's most basic form of arranging deals with objects in real space." And in reviewing Polanyi (p.104) they state that "central to Polanyi's argument is the belief that all knowledge is ultimately personal, rooted in and affected by ways of believing and valuing that the knower has internalized."

In reading the various attempts which Robinson and Petchenik have made to define the map – and particularly the role of one who maps, the mapper – it is not difficult to extend their discussion to see the action of mapping as being a form of improvization. It is not clear, however, that this was part of their intention, for they define mapping in this way (p.44): " 'Mapping', although normally a participle, will be used here as a gerund to denote the entire system with its own structure and processes, including the concepts of mapper and percipient, as defined in Chapter 1." But perhaps it is the process of thinking about a specific geographic phenomenon, of seeking out its most characteristic dimensions, of setting out communication goals, and of considering the possible modes and forms of its graphic representation which are of more fundamental educational value than the mechanistic execution of the final product. In other words, mapping (as defined here) may be the more essential experience on which to base our approach to education in cartography and perhaps even geography. If this is so, then our present approaches to cartographic and graphic education (and ultimately geographic education) may be faulty.

A survey of materials intended to introduce maps to children (Castner, 1984b) suggests that they concentrate overwhelmingly on a very narrow range of traditional components of maps – their conventional symbols, their expression of scale, their projection system and grid, and their statement of orientation. The method of introducing "map skills" stands in direct contrast to the two music education methods discussed above in that the child is first introduced to the joys of music through participation and "performance" in the sense of completion and sharing of a piece of music. Written notation enters the picture only when the child is ready to appreciate the need for writing down what is performed. Even then, it is not introduced as a hurdle which must be jumped over before further progress is made, but rather as a corollary skill that can be learned at the same time, but not necessarily at the same rate, as the performance skills. The Suzuki approach looks at this manner of introducing written notation in a very useful way by asking: "When is a child able to read every word he speaks?"

The Orff approach is much more specific about this: "We must see that the child experiences music and creates music *before* he is told how it is put together" (Hall, 1960, 18). In essence this method begins with the voice, traditional rhymes; and songs; and adds simple movement,

such as body percussion and clapping; and locomotion, such as walking and dancing. Then, and only then, are the rhythms transferred to the
instruments and true, primitive ensemble music is made. "When he is
completely at home with the medium and is capable of expressing
himself, musical notation is finally introduced" (Hall, 1960, 18).

The implications for cartographic education are obvious. *Map making*
and *mapping* are two quite separate procedures. To date it would appear
that our traditional concerns have been focused upon the former, the
map and its structural components, rather than on the latter, what we
might wish to communicate through the process of mapping. A Suzuki
or Orff type of approach would suggest concentrating more on the
latter, on an active mapping program in which a limited number of
graphic elements are manipulated, and from which a useful variety of
both map products and expressions of geographic information may
emerge. In this regard, Petchenik (1979, 10-11) makes the interesting
distinction between two essentially different kinds of map meaning,
between two distinct forms of knowledge: "On the one hand, there is
the map whose meaning lies in the human experience of *being-in-place*;
this has come to be thought of as the general reference map, and its
meaning can be considered cognitively less complex. On the other
hand, there is the map whose meaning is experienced as *knowing-about-
space*, and it is this situation that is most commonly associated with the
thematic map."

From this it could be said that our traditional educational approach
has been directed toward "being-in-place" knowledge as expressed in
maps as reference documents or storehouses of information. Such an
approach may have led us to have overly concerned ourselves with
emphasizing the various conventions in maps, such as the conventional symbols that, by and large, resemble the objects they stand for
(are similar in their pictorial attributes). Perhaps the action of mapping
offers an alternative educational approach. Through thematic mapping, especially at very large scales, we can consider the more abstract
or collective attributes of things which we map. In so doing, the information being presented takes on importance greater than the elements
used to represent it, although they can be seen in terms of how they
collectively work together to provide a common "framework of expectation" (Rosenfeld, 1985) that allows us to recognize some of the essential forms that graphic displays can take. In such a context, the various
structural dimensions of the symbols (their size, shape, color, or area)
become more significant than their pictorial attributes.

The structural dimensions of symbols suggest another analogy with
sound and light which in turn may suggest some more specific directions in which to discover new ways to address old cartographic topics.

The four basic attributes of sound are pitch, loudness, timbre, and interval. The similar and corresponding attributes of light[3] are hue (wave length), value (brightness), chroma (purity or intensity), and texture (pattern). An understanding, appreciation, and ability to discriminate variations in these visual dimensions is fundamental to developing one's visual perception. These abilities may also foster the student's capacity to react to and process all kinds of visual stimuli, including the graphic displays about which we are concerned – photographs, video screens, maps, and the like.

More specific emphasis might also be placed on (1) the process of mapping rather than on the map itself, (2) allowing children to discover and experience, through some kind of improvization, the relationships among various elemental perceptual principles in vision, and (3) setting the core activities of mapping in a broader educational context. This third point would mean making more obvious the interrelationships between mapping and drawing, visual perception, geography, mathematics, and geometry. By focusing some of our energy on improving children's skills in visual perception, we may not only improve their abilities to cope with the complexity of map images, no matter how they are produced, but also eventually reduce the significance of the great variation in map user experience that presently constrains our map design research.

The following explores other connections between music and graphic expression that are worth mentioning and which may also provide ways of incorporating some of these ideas into our teaching of cartography and geography.

Rhythm in Music

The combination of sounds that create music can be characterized by the three elements: rhythm, melody, and harmony. Of these, rhythm has apparently received less attention from theorists despite being indisputably the most basic element. Rhythm has both an obvious temporal and a spatial dimension. The great Swiss music educator Emile Jaques-Dalcroze,[4] noted the interrelationships between the muscular energy deployed, the space to be traversed, and the time in which a movement takes place. Thus rhythm is essentially physical, it is movement. The system of rhythmic education which he developed is known the world over as *eurhythmics*; it cultivates the child's rhythmic potential through the medium of his own body. One of the objectives of eurhythmics is to develop habits of listening in which children learn to identify what they hear with what they do (Findlay, 1971, 1-2).

The temporal dimension of rhythm can also be explored through language as in any metred poem or measured rhyme. Spatial aspects of rhythm are experienced by the muscular feedback of movement, a self-communication if you wish. Here, one's presence in or awareness of a particular place is worked out through the experience of motion within it: the obvious educational goal is to help the child articulate this awareness in some way.[5] Certainly the spatial aspects of rhythm can be represented in some graphic way, namely, mapped.

First, there is the very obvious use of musical notation to *graph* the various structures within music (within the "contour of the melody") (Pick, 1979, 146). This includes patterns of relative and absolute interval sizes between successive pitches (and their rhythmic relationships) and within them, that is, the structure which produces the actual harmonic arrangement of chords.[6] There is a useful parallel here with our teaching about maps. We seem to be more intent on teaching about the elements of maps (the notes) than on the information that they represent (the intervals or relationships between the notes).

Secondly, the more relevant activity concerns the movements in space itself as set to rhythm and melody. Depending upon the words themselves, there can be a more or less explicit movement in real space. Movement activities in the elementary grades may be the child's first formal experiences in verbalizing and acting out relationships, in both personal and relative space, among a number of nearby objects or places. Activities could include those which emphasize aspects of direction, orientation, and measured distance as the focus of change in a progression of movements. In addition, movements could be elicited from the children which reflect the textural dimensions of patterns that can be obseved in various kinds of graphic images (pictures and collages) or textured surfaces (cloth and walls). The potential for variety and complexity in dealing with these few concepts can be seen in the number of ways in which movements can take place. Birkenshaw (1986) describes the elemental building blocks of movement as: "balancing, walking, running, sliding, and leaping come first, and next, when the child is able to have both feet off the ground at once, jumping and hopping. Finally comes galloping and skipping. In addition, of course, we have the non-locomotor movements such as swaying, stretching, rising, clapping, snapping, patting the thighs (*patschen*) and (especially for nonambulatory children) tapping with one hand, nodding the head, etc." When we consider this rich collection of elements of motive expression in light of the second fundamental principle of the Suzuki method (that every step be precisely identified and mastered), it is clear that any identification of basic concepts and principles is a far more complicated undertaking than we might first imagine.

In any case, the representation of movements in space presents us

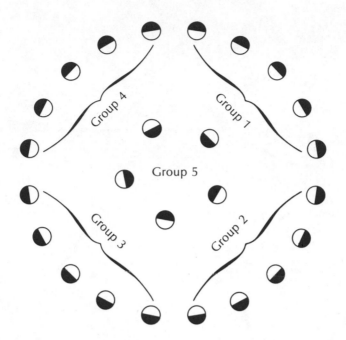

Figure 20 The initial formation from which a Lugenmärchen is to be performed by five groups of school children. After Weikart (1981,16).

with another opportunity to develop a method of map making. At the adult level, we have very sophisticated systems for representing the choreography of classical ballet. But obviously we would not try to use these complex images and systems in elementary school classes. There are, however, graphic techniques for representing both the arrange= ments for initiating an activity in dance or movement (figure 20), or the resultant patterns that are to be followed, as in figure 21.[7]

Here is a communication problem that provides all the ingredients for learning a new language, albeit a graphic one. There is a need to record a procedure so that others, in another place and at another time, can re-enact the procedure. As Bruner notes (1978, 45), language learning involves solving problems by communicating in a dialogue, and that dialogue occurs in a context. An additional advantage of attempting to map a dance or movement routine is that there are no previous solutions or conventions from the cartographic literature. As a result, the child and the teacher can start afresh and truly explore all the variables in graphic expression so as to arrive at symbols and designs that suit them and the tools and materials available to them. The only

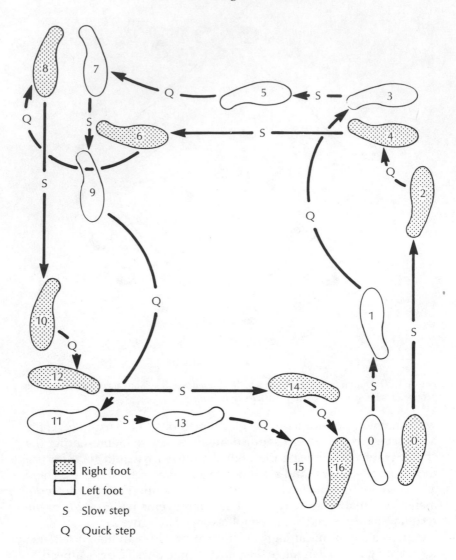

Figure 21 The pattern of steps taken in the fox trot, step V. After Hall (1963,200).

test of their success is to give the "map" to another class and have the students reproduce the movements of the dance.

Music and Geography

Finally, there are distinctive sounds that, for various reasons, we come to associate with various regions of the world. The folk songs, dances, and ecclesiastical music of many countries have unique rhythms, tonalities, melodies and instrumentations which we learn to associate specifically with those countries. Many of these folk songs and dances can be appreciated and performed in music classes; they also can be heard as orchestral works in which more complex musical textures have been developed.

By and large, it is probably aspects of cultural rather than physical geography which provide us with the strongest musical associations. However, there are also pieces that reinforce ideas about the physical landscape. For example, much of the symphonic work of Jean Sibelius (Finland) seems to convey images of a bleak, rock-bound winter landscape. Debussy and Vaughan Williams both wrote highly descriptive pieces about the sea. Villa-Lobos has tried to capture something of the exotic nature of the Amazon River and the Brazilian rainforest.

For any individual country there are often three categories of music: those, generally, of the indigenous composers, place-specific pieces, and those composed by visitors who came away with their own impressions of what they saw and heard. In the first category one could name for the USSR, Tchaikovsky, Rachmaninoff, Prokofiev; for France, Debussy, Bizet; for England, Vaughan Williams, Britten; for the USA, Copland, Grofe, and so on. In the second category, there are pieces such as "Fingal's Cave" (Mendelssohn), "Great Gate of Kiev" (Mussorgsky), "Alexander Nevsky," the battle on Lake Peipus (Prokofiev), and so on. In the last category, one can think of Dvorak's "New World Symphony," Tchaikovsky's "Capriccio italien," Chabrier's "Espana," and Brahms' Hungarian Dances.

Thus, in studying the geography of a particular country or region, there are obvious listening options or choices. If we are sincere in attempting to present and understand the full flavour of a geographic place, then we should be attempting to provide students with some opportunities to hear the sounds of that place, in this case the sounds of their music. In chapter 8 I will mention the other sounds of geographic places – the sounds of the indigenous spoken language.

CONCLUSION

This examination of music, and especially of two of the methods in music education, provides a number of philosophical ideas about how we could provide children with experiences that might improve their basic skills in visual perception and spatial awareness. These ideas revolve around defining a visual analogue to improvization; it may be that mapping, as defined here, will serve that purpose.

Implicit in the application of an improvizational approach to geographic education is the creation of spatially and graphically stated questions for which there are more than one answer. But these answers will be constrained by a number of conditions or options so that there is a relatively small number of reasonable answers, some better than others. What is more, the evaluation of "better" or "worse" should be apparent to all concerned (to both student and teacher), and should not simply be a judgment made by the teacher. As I suggested in chapter 1, the creation of a set of answers which may vary in its quality provides a context in which the original question can be posed again (1) in a slightly different light; (2) at some greater degree of complexity; or (3) at some other level of generalization. As a result, the core of the educational curriculum becomes spiralled instead of sequenced. It is the skills of analysis that should become sequenced and not the core concepts of the discipline.

To accomplish this, however, it is imperative that the component steps involved be precisely identified and elaborated. We have done this for map making, narrowly defined, but not for mapping and spatial awareness, broadly defined.

This shortcoming is evidenced by the striking contrast in our use of formal map notation. Geographic instruction usually begins with the mastery of the conventional symbology of maps and the search for particular symbols on maps. This type of learning often precedes opportunities to experience first hand the relationships between objects in space and aspects of movement in space. Rarely are children allowed then to consider the possibilities for the graphic expression of these kinds of information. In other words, our approach to geography does not aim at participation and performance. In order to change this, we need to develop activities and map forms that do not depend initially on a "formal notation." Rather, as a prelude to introducing the useful cartographic conventions, we should be working with the more general aspects of graphic expression. Eastman perhaps put it best when he declared (1985, 100) that map reading is a creative process involving the meeting of two organizational processes – those of the map maker and those of the map reader. To make our map designs creative, we need a set of 'tools' for map creation, not a set of 'rules' for map construction.

It may also be that we should consider some serious collaborations with elementary level music teachers, for it is they, who in using movement as a basic component of their teaching, are actually doing some of the fundamental teaching in spatial awareness in areas we might have regarded as geographic domains – direction, orientation, pattern, and measured distance (scale).

CHAPTER FIVE

A Connection with Art

Only what I can imagine completely and can redraw from imagination, only that have I truly understood.

Goethe

'Creative activities' are scheduled nowadays in many classrooms, but they do not lead to creative thought ... The superficial use of creative projects does more harm than good, since it denies the basic seriousness of the endeavor. There must be a definite problem with definite limits for which preparatory training has been so complete that the children can be allowed to solve the problem almost entirely by themselves.

Isabel Carley (1977b)

Capitulation to the view that arts education is not school's responsibility will remove from the school what it needs most: activities and problems that stimulate the imagination, tasks that celebrate ambiguity, prize sensitivity, and encourage children to take intellectual risks.

Elliot Eisner (1980)

Implicitly, art would seem to be an area from which cartography should draw a great deal of creative inspiration and technical knowledge. After all, both areas are involved in one way or another with the representation of our circumstances here on earth. Indeed, cartographers have usually considered their discipline to be a meeting place between science and art (see, for example, Robinson et al., 1978, 6).

However, the connections with art are neither clearly identified by practitioners nor recognized by educators. Perhaps the stated goals of art education are too divergent from those of cartography and geography, or the media of expression and range of emotions conveyed are too diverse to allow a common ground to develop. Yet clearly visual perception is involved in both areas. As well, they both have the desire to serve or influence viewers of their products (although this may be less true of works of art). This desire is facilitated by both the subject matter chosen and by the various arrangements and treatments of the graphic components making up their products. Finally, both areas are concerned with the interactive relationships between thinking, seeing, and graphic production. Surely there is sufficient commonality here to allow for some cooperative approaches to education in the elementary and intermediate grades.

The basic goal of this chapter is to examine (1) some of the possible reasons why such linkages have not previously been made, (2) some theoretical ways in which the disciplines can be seen to be closely related, and (3) some suggestions as to practical areas in which common goals of cartographic, geographic, and art education might be advanced.

CONFESSIONAL

There should also be a place in this introduction to admit to the great difficulty I experienced in writing this chapter. It relates very much to my mixed emotions about art. Perhaps by declaring my bias at the outset, both the reader and I can take a more positive view of what follows.

My problem is one of strong, conflicting personal feelings that arise out of a certain amount of success, on the one hand, in particular kinds of "artistic endeavours" (map making, graphic design, sandcastles) and, on the other, a certain amount of uneasiness in the company of much modern art and of products that I see coming from school art classes. I am humbled by my own efforts in drawing some kinds of things, especially hands and faces. As a result, I am well aware of the limits of my talents. But, in contrast, I can draw objects and animals fairly well, because I know what attributes I can work with that convey, for example, "catness," and I am not afraid to exaggerate these features. Thus I am very much aware of a cognitive and a perceptual side of drawing.

Most of my conscious effort employed in a creative activity involves making use of concepts and experiences learned from cartographic design, and not from courses in art. I believe this to be so because much of the effort in cartographic communication research has been directed at explaining the "scientific" underpinnings of much that we thought was "artistic" in map making. If we are to grow as a "craft industry" (in the old sense of the craft guilds), then we must be more articulate about the artistic decisions we make, namely, how we make them, why we think they should produce the desired effect, and so on.

To date, many artistic decisions that cartographers have tended to make have been shown to be quite good, in that they demonstrated good artistic sense, if you will. This would suggest that all that research into map design was not worth the effort. But without it we would not know why those decisions have been so good. This is a much firmer ground on which to pass on collective wisdom. Thus I am convinced that the rules of thumb and perceptual truths that have come out of cartographic design research and teaching are effective in helping me to amplify my modest artistic talents and to make everyday use of various informal forms of graphic thinking.

It is regrettable that I did not come to understand these principles earlier in life. But I think that, because of their value, they should be more visible in the school curriculum. Some of the effort behind the writing of this chapter is directed at rationalizing a place for them in the schools.

THE NATURE OF ART

When we think generally of art, a number of aspects come immediately to mind: its role in self-expression, its technical side, its history, its relation to aesthetics, and its place in design. But, stated in these ways, the relationship between art and cartography is not so easy to establish. Perhaps it is more useful to think of art in its educational realm. There it has been described in this way: "The process of drawing, painting, or constructing is a complex one in which the child brings together diverse elements of his experience to make a new and meaningful whole. In the process of selecting, interpreting, and reforming these elements, he has given us more than a picture or a sculpture; he has given us a part of himself; how he thinks, how he feels, and how he sees" (Lowenfeld and Brittain, 1982, 3). Stated in this way, two crucial factors come into focus: the child's experience and his manipulative activities.

The Experience Factor

The experience factor is a primary one in differentiating what one child, as opposed to another, may do in a classroom situation. It obviously contributes to the great variation in the ways children will generate, meet, and solve problems, as figure 22 suggests. But what experiences are we talking about? Two possibilities come to mind: experiences in doing things and experiences in looking.

Children have experiences in doing things, both at home and at school, with family as well as friends, which provide them with knowledge about the world. These experiences furnish the children with subject matter when they are called upon to contribute to school activities. This should be seen in most classrooms as a positive factor, one to

Figure 22 *Classroom Charades* by Charles E. Martin. A skeptical view of nonverbal communication. Courtesy of *The New Yorker*. Also reproduced in Gombich (1960).

be taken advantage of especially when a diversity of ideas is sought. It may be, however, that it will be seen as a negative factor when the goal of the classroom activity is for everyone to do something that not every child will be able to perform successfully because all experiences are not equivalent. This is a condition that the educational system tries to neutralize over a long period of time, but it is something the classroom teacher must adapt to on a daily basis.

This leads to the second possibility: that children have had experiences in observing things, namely colors, textures, surfaces, spatial arrangements, things foreshortened, things occluded, and so on. While the subject matter of their experiences will be quite diverse, the nature of their visual experiences, expressed in these dimensions, should be relatively uniform. Thus these latter kinds of experiences would seem to provide a common ground on which education in visual perception could begin.

In neither case will these two kinds of experiences provide children with anything that would allow them to test whether or not a graphic image had any particular artistic merit. This is a special problem that we will return to in the next section of this chapter.

The Manipulation Factor

The manipulation factor relates to what the child "does," both in terms of objects and processes. The objects utilized in an artistic activity could obviously be both physical and intellectual, (in the mind's eye). In one typical early activity in art, one that is unfortunately all too common, a teacher may supply the children with identical sets of materials and instructions. It is easy to spot the products of such activities at a school open house: walls covered with finger paintings, pumpkins, or other seasonal images, caterpillars (the classic egg carton and pipe cleaner creations), and so on. Each is individualized by the child's ability to letter, to cut paper cleanly, or to paste carefully. All products are successful, but they provide few opportunities for intellectual inputs nor do they lead to creative thought.

Certainly there are important motor and hand-to-eye coordination skills being developed in these kinds of activities. But they fail dramatically to provide the rich perceptual experiences that a child would receive in, for example, an Orff lesson with a combination of singing, movement, and body percussion. This begs the question: what art activities can also contribute to the development of the child's perceptual skills in vision? Or more specifically, what activities in art education can provide the same perceptually rich experiences in manipulating the principal perceptual dimensions of light as Orff lessons provide in manipulating sound?

I have not been very successful in uncovering an answer to this question from the teaching materials and art books that I have examined. Salome, twenty years ago (1965-6, 18), referred to the "current reemphases upon the contribution of art education to the development of visual perception," and that "art education objectives usually include perceptual growth as an important goal." However, he notes that "with one exception, recent art education texts and curriculum guides either fail to mention or include information pertaining to perceptual training." In fact, he declares that there is no empirical evidence to support the assumption that visual perception improves as a result of participation in art activities! Seven years later (Salome and Reeves, 1972) he drew essentially the same conclusion.

But what about the other side of the coin? How might art products (and indeed other graphic products) improve as a result of children's participation in activities designed to sharpen their powers of visual discrimination and perception? It is clear that the primary perceptual dimensions of vision are not articulated in art education in the way aural skills are developed in music education materials. If they were, we would see more activities that encourage their formal manipulation, that is to say improvization with them. Surely the development of a keener sense of what these dimensions are, an ability to discriminate among them, and confidence in working with them are worthy goals. They lay the groundwork for so many other activities, both within the world of art and in many neighboring disciplines as well.

Perhaps it is through the introduction of more specific graphic

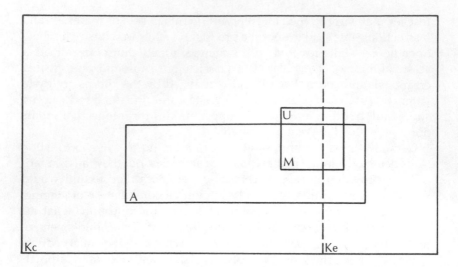

Figure 23 The schematic creation of a graphic product or image. See text for a discussion of this and the following diagrams.

problems and the need to communicate their solutions that we can inject a measure of graphic improvization into classrooms. This may be a more useful ground on which to explore the place and role of educational experiences in visual perception.

THE MATTER OF COMMUNICATION

While it is possible that certain goals and attitudes have kept instruction in art and cartography quite separate, there are numerous contexts in which they can be seen on common ground. In their discussion of the map as a communication system, Robinson and Petchenik (1975, 10) provide a useful diagram of the significant cognitive fractions which are involved in communication. I will describe the diagram in the broad context of graphic communication. Start with a large rectangle (see figure 23) representing the total conception of knowledge about the world held by mankind. The vertical dashed line divides the conception of the world into error-free (Kc) and erroneous (Ke) segments. Within the overall domain of man's knowledge, we can imagine any number of smaller rectangles (such as A) representing subsets of that knowledge held by individuals, whether they be map makers or artists. Some may be larger, others smaller. Some may be further to the left, others further to the right, although none lie exclusively on one side of the dashed line.

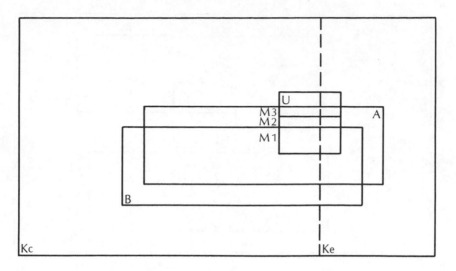

Figure 24 The significant knowledge fractions involved in graphic communication.
After Robinson and Petchenik (1975,10).

Within any given culture, there would be considerable overlap among
a set of such rectangles because each culture holds many ideas, attitudes, and pieces of knowledge in common, and which in themselves
form a basis for defining that group. As Rosenfeld (1985) puts it, the
overlap provide the rules for the common "framework of expectation"
between the artists and viewers which allow communication to take
place.

If we superimpose a small square onto the diagram to represent a
graphic product or image of some kind, it will overlap to a considerable
degree with its creator's knowledge realm (M) because that is its origin.
It is not unreasonable to assume, however, that most graphic products
will have some component of error arising out of an incorrect conception held about the world, the area of the square to the right of the
dashed line. There is probably another component of the graphic image
(U) outside of the creator's subset of knowledge, which represents a
potential gain of knowledge resulting from either the production or
subsequent viewing (or both) of the image by the creator himself.

Once the image is made and is given to someone to examine or use,
the diagram, figure 24, must now include a second rectangle (B) – the
conception of the world of someone viewing the image. Again we can
imagine a whole array of possible rectangles representing all kinds of
potential viewers. In order for there to be some kind of communication,
the two rectangles and the square must overlap to a certain degree.

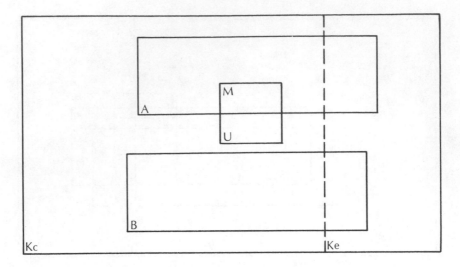

Figure 25 The knowledge fractions involved in the creation of a piece of nonobjective art, namely a graphic image without any precise communication intent.

Unless the creator and the viewer share some knowledge, both about the world and about conventions of graphic communication, then there is no basis or context in which other information can be placed.

Robinson and Petchenik are very precise about the components of this interaction. In figure 24, M1 represents that part of the image information that was known by both parties. M2 symbolizes that part of the image's information not previously known to the viewers (the map percipient) and thus constitutes a direct increase in their understanding and knowledge. For cartographers, this is often the most important component, the one for which the map was made. A third portion of the square, M3, represents that part of the image information (also not previously known to the viewer) not comprehended by the viewer; it illustrates a discrepancy or breakdown in the communication. The fourth area, U, stands for the increase in knowledge or understanding that may occur from viewing the image, but which was neither intended nor accounted for by the creator.

In cartographic communication theory, it is felt that the success of a map communication can be measured in a number of ways and that its success can be thwarted by an overabundance of detail (noise) at various places in the communication system. Presumably, we share a certain concern with artists about the success of our graphic products. However, the information carried by maps is often more obvious and specific; the emotive qualities of a work of art, on the other hand, may be of greater importance to the artist. Perhaps the following diagrams

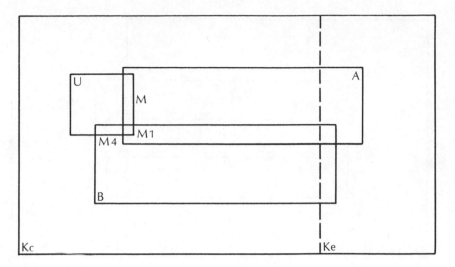

Figure 26 The more likely knowledge fractions involved in the creation of a work of art with a significant component supplied by the viewer.

will clarify the difference.

At the lowest level of communication, an artist can produce an object of art for viewing and can declare that its meaning "speaks for itself." He can reinforce this declaration by giving the work no title or a very ambiguous one, by not describing it in any way, or by excluding from it any recognizable objects, symbols, or apparently meaningful elements. In the extreme case of an artist working in one culture and a viewer coming from another, there would be no overlap between the square, (the work of art) and the viewer's rectangle, as in figure 25. The work would ostensibly have little meaning in experience to either party, although the areas M and U remain potentially for the artist. It would be difficult to consider the work of art to represent an erroneous aspect of knowledge, so the square has been placed to the left of the dashed line. While we all have seen "works of art" which seem to fit this example, it is probably an extremely unlikely occurrence.

A more likely representation, as in figure 26, would involve a small amount of overlap of the rectangles (even though neither party might recognize it) producing areas M, M1, and M4. This latter area would represent that component of the image which appears to be meaningful to the viewer but which cannot be attributed to anything intended by the artist. Area U is undefined because that is what the artist wishes the viewer to provide his own interpretation; it is probably the largest component, for that is the way the artist wishes the "communication" to take place. In other words, in such a situation, the work of art serves simply

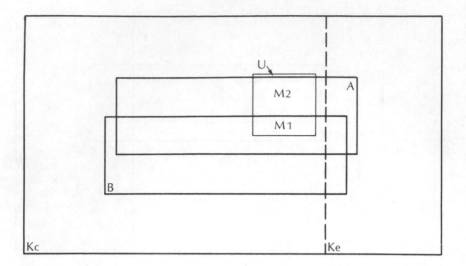

Figure 27 The knowledge fractions involved in the communication with an architectural or engineering plan.

as a stimulus for reflection. Any meaning assigned to it must come from the viewer. The amount forthcoming will probably be a function of the time allotted for that reflection, the sophistication or intellectual imagination of the viewer, or the viewer's ability to verbalize his thoughts about the work of art or elements of it.

At this level, there is no communication in the sense of figure 24, but rather only a stimulus-response situation in which the former is the responsibility of the artist, the latter is that of the viewer, and the two processes are independent of one another.

At the other extreme, a work of art can be appreciated for the accuracy with which it replicates a real world object or scene. Numerous examples come to mind of images which are technically of high achievement because of their detailed fidelity to the original. Of these, perhaps the illusions of *trompe l'oeil* (Battersby, 1974; Frankenstein, 1970) are the highest expression of realism. However, when we consider faithful reproduction, we more often think of photographs, architectural drawings, and various engineering plans. These record exact and accurate aspects of constructs so that they can be manufactured or assembled by people other than those who produced the drawings. Thus for many of these purposes, the variation of scale across such drawings must be controlled or made predictable in some systematic way so that measurements can be made from them.

To represent this situation in a diagram, figure 27, the square for the plan or blueprint would presumably also have only two main divisions, but this time M1 and M2. By using a conventional format, projection system, and symbolization that are mutually understood by both parties, the designer and the manufacturer, a maximum amount of accurate information can be conveyed. The area of M3 should be virtually nonexistent; that of U might still be present but would be insignificant in terms of the overall goals of such a communication. One would also assume that the square is located exclusively to the left of the dashed line, particularly for things like bridges and skyscrapers!

Between these two extremes lie the bulk of man's artistic effort and probably most of our cartographic production. This situation is represented in figure 24 where all four components of the square are present. For maps we presumably wish to make M2 as large as possible and M3 and U as small as possible. However, we would hope, for the sake of variety and the joys of serendipity, that U is not eliminated completely. For some time now we have realized that M2 cannot be achieved by simply increasing the amount of information placed on the map – something our improved technologies have made possible. But rather, we can very quickly increase the map's content until we have reached an overload situation with our viewers – the map becomes simply too complex and cluttered (noisy), or imposing. This problem is illustrated in figure 49, p.157. The question arises: What knowledge of principles of visual perception will allow us to best control the amount of noise in our graphics so that the positive aspects of our communications can more easily be seen? By clarifying the perceptual processes operating, we establish the viewer as an important component in the communication equation – something which may be missing in art education. We should consider this point in some greater depth.

ART EDUCATION AND COMMUNICATION

I have suggested that an intuitive relationship exists between art and cartography because both are graphic modes of communication. However, evidence from North America would suggest that the thrust of art education is less concerned with communication, as I described it in the previous section, than with teaching self-expression, the manipulation of various media, and the development of an appreciation of aesthetics. For example, the following statement comes from a position paper published by NAEA (n.d.):

Through the ages, man has used the arts to build and enrich his personal and shared environments and they should lead to the desire and the ability to

improve them. Any art education program which consistently emphasizes the
ability to make qualitative judgments can help each citizen to assume his share
of responsibility for the improvement of the aesthetic dimension of personal and
community living. Acceptance of this responsibility is particularly important
during periods of rapid technological change.

Only recently have we, as map makers, become aware of the tremendous importance of the map user or map percipient in determining
success, however defined, in the map communication equation. This
would seem to have a direct parallel to an artist who is intent upon
conveying an idea or an observation about his culture through his work
of art. He must first consider the various signs, symbols, or codes which
carry particular cultural or perceptual meaning to his intended viewers,
and how, through his graphic design, he can order his graphic elements
so that the viewer can gain some appreciation of the cultural idea which
he wishes to communicate. Without this consideration of his audience,
the chances of their perceiving his message are greatly diminished and
his work becomes a much more limited exercise in self expression.
   On the other side of the coin, however, audiences with an understanding of the ways in which ideas can be represented graphically, and
how graphic elements can be arranged or manipulated are in a better
position to be able to analyse what the artist has done and is trying to
say. Then, in turn, these nonartists can utilize in their own modest ways
these graphic devices in their own initial and unsophisticated attempts
to communicate ideas.
   Thus for there to be any communication between artists or map
makers and their audiences, there must be some aspects of the images
that are commonly understood by both parties, – producers and users.
Usually, the thematic content is an important and obvious component
of this mutual awareness. But there are also ways in which color,
contrast, texture, form, and other primitive dimensions of images can be
utilized to add emphasis, direct attention, make statements about
accuracy or authenticity, and these can be used both in maps and in
works of art. Thus the ways in which these variables can be manipulated
to produce various perceptual effects should be part of the common and
introductory training of anyone who wishes to make graphic images to
communicate information or ideas. I will provide a brief example of this
presently with Leonardo's famous painting of *The Last Supper*.
   The question arises as to where this kind of fundamental information
has been and should be taught. It seems clearly to be a topic at the heart
of graphic communication and visual perception and thus a logical
topic for elementary art education. However, if the specificity of communication is not an educational concern, and it does not appear to be

in most early art experiences, then it may fall to others who are involved
to teach concepts that are fundamental to communication. Thus, one
finds many of these concepts addressed in introductory textbooks on
cartography because students have not received them heretofore and
need them in order to increase their understanding of the ways, and to
what effects map images can be manipulated. The last two decades have
seen a great deal of exploration of the principles that underlie the
perception of most all of the basic symbolic elements which we use in
maps, whether they are to be seen as figure or as ground. The question
remains whether this effort should continue and where in the curriculum students should be discovering these principles.

THE PROBLEM OF CREATIVITY

Bruno Bettleheim (1980,424) raises a question that is a source of some
confusion – the problem of creativity:

Nearly the whole of the student's school experience relates uniquely and
exclusively to the recorded experience of others. Most of the problems have
ready-made answers in the back of the book or, more up to date, in the preset
teaching machines. Our educational system is not concerned with originality or
creativity; it is concerned mainly with acquiring a body of knowledge narrowly
defined; with the memorization of facts, the finding of ready answers to
problems, answers that are already known to somebody – or the test could not
be scored.

Art, he goes on to contrast (p.425), is the only subject in the educational
experience where members of the future generation can be offered the
chance to truly find themselves as unique individuals. Only in art are
there no ready-made answers telling students what they ought to see,
feel, and think. We have a problem when either art teachers insist that
they know what is good or bad art or they uncritically accept any artistic
product as being "imbued with the artistic spirit."
    What a burden this must place upon an art teacher! Make a judgment
and you have crushed the child's creativity; if you are uncritical or vacu-
ously praising, children receive an equally false impression of their self-
worth or artistic skill. In both cases, the teacher's reaction rests upon
some criteria or value to which the child is not a party.
    Bettleheim (1980, 413) also chides art teachers for not recognizing in
their own creative efforts the tremendous amount of discipline which is
necessary to achieve a significant work of art. The nature of creativity he
declares (1980,414) "is not any unconscious outpouring, but the process
whereby carefully selected and arranged elements of such fantasies are

rigidly worked over by a critical mind in a most disciplined way within the framework of a well-understood tradition." Bettleheim, however, makes it clear that the notion that everybody can paint is not the same as the idea that everybody is an artist (1980, 421). He gives an example in music of a Bach piano piece. While thousands can fumblingly play it, few would delude themselves into thinking they are great musicians. But perhaps their noncreative and nonartistic efforts provide them with a deeper appreciation of what Bach has written, a greater awareness of the achievements of truly gifted pianists, and a heightened trust in their own aesthetic experience of a great work of art.

The Confusion Between Artist and Artistic

Part of the educational difficulty may lie in the confusion between the states of being an artist and being artistic. The former involves gaining considerable experience in trying to solve new problems of graphic representation and symbolism, achieving a high level of technical skill in some media of expression, and reaching a point where one's works are seen to transcend themselves and speak to universal principles of human thought and being. These achievements usually require years, if not decades, of maturation. Reaching the status of being an artist involves a great deal of innate talent, probably some excellent training along the way, and many years of hard work. Given the small number of people who can have successful careers as artists, the common curriculum should not be addressed to their particular needs, any more than our first courses in cartography should cater to future professional map makers rather than to the host of potential users of maps.

Being artistic, on the other hand, is a more general concept that implies a certain amount of skill in the execution of simple drawings or arrangements and in the attractive assembly of particular elements or objects. It also involves some evidence of those intangibles of taste or aesthetic appreciation and perhaps some ability to verbalize their role in these activities. "Artistic" does not carry élitist connotations to the degree that "artist" can. In its most simplistic expression, being artistic implies having "an eye" for things like neatness, order, and harmony and "a hand" for clarity or beauty in things like writing, sketching, or arranging.

It would seem that many of the attributes that we associate with art in a negative sense are those that relate to being an artist: the occasional arrogance of modern art when it seems to be pulling our legs; what one critic describes as art having been circumscribed and subverted until it serves only a blind self-interest (Argüelles,1975,2); the sometimes lofty claims made for works of art; the often trivial products that pass for art or which children bring home from art classes and so on. Being artistic does not seem to carry quite as many negative associations. While many people consider not being artistic one of their personal ineptitudes (Papart, 1980, 8) and will try to avoid being drawn into a situation which may require a demonstration of their artistic skills, these same people clearly do make judgments and decisions in their lives based on their visual assessments of everyday realty– for example they choose clothes, household furnishings, cars and so on. Therefore, they may have just as many opportunities as artists do to exercise skills in visual judgment. The nonartists, then, should also have some training in learning how to look with discrimination to help them in these judgments. As Broudy notes (1983, 234), it is far more reasonable to expect an educated layperson to perceive in the manner of the artist than to perform like one.

It is not my intention here to state the case for art in the school curriculum – a topic of great complexity and debate in itself (Smith,1978). But rather it is my intention to argue that we should be providing learning experiences involving the basic processes of visual perception. This would serve three purposes. First, it would avoid any of the negative and élitist aspects of art that may be a part of training in the first six or eight school grades and which might be directed more at producing artists. Second, it would also discourage teachers from falling into the trap of making themselves the sole judges in their classrooms of what is or is not artistic. Third, by working toward the improvement of children's visual powers of discrimination and analysis, we contribute to both those who will become artists and those who, we hope, will retain or gain confidence in their abilities to be artistic. In the long run, of course, a large, articulate population on artistic matters will be better able to support and appreciate the work of the artists and graphic designers in our communities.

Thus the problem for the art teacher, as Bettleheim sees it, is the potential confusion between the goal of teaching painting as a leisure-time activity for the masses and that of providing a way of lifting society out of the humdrum of mass living and of challenging individuals to strive for a higher integration of themselves and of our society. He notes how art educators (1980, 422), in writing for their colleagues, make fervent assertions of how practical and rational their discipline and how it contributes to a more comfortable and better life. There is no doubt, however, how he views this problem, for he declares: "We shall have to give up the notion that art can be everything to everybody, because if it is, then it adds up to being nothing of real importance to anyone."

I find this too severe a stance for it fails to explore the middle ground that encompasses both visual perception and graphic communication – areas in which we need not become concerned with the abstract

Figure 28 Leonardo da Vinci's painting of *The Last Supper*. Refectory of Chiesa di Santa Maria delle Grazie, Milano. By permission of Alinari/Art Resource, New York.

questions of artistic worth. If art educators do not feel these areas are compatible with their goals, then it may be up to geography teachers to provide those bridging activities which will allow students to develop both their powers of discrimination and an understanding of graphic communication without developing a fear of "not being an artist."

COMMUNICATION THROUGH ART

The practised "Master" will have worked on his ideas for such a long time, in so many variations, that he has developed an intuitive approach to his work. But at some time in the past he must have consciously struggled with an idea and how best to express it in his work. From an educational point of view, it is this latter struggle which would seem to be more rewarding a subject for study. In order to communicate an idea, the artist must consider the various symbols, signs, and codes which carry particular cultural meaning and how, through his graphic design, he can order his image components so that the viewer can gain some appreciation of the cultural idea that he is attempting to present. This, of course, is the essence of "mapping" as I have defined it in chapter 4.

Figure 29 The six ABCs of linear expression. After Watkins (1946,10). Courtesy of the Phillips Collection, Washington, DC.

In both cases, it is the struggle that is of educational value. Teachers of both art and mapping face the same trap of allowing emphasis to shift away from the struggle, the process, to the product of the struggle. Thus an analysis of how various artists have produced contrasts and arranged elements to focus attention on certain parts of their work would be a component of art education useful to both art and cartographic education, because it sheds light on that communication struggle.

However, Bettleheim questions courses (1980, 425) that tell us how to look at pictures or what to see in them. He despairs that the great poets were for decades ruined for him because he had been taught what made them great and how he should enjoy and understand them. It seems to me to be one thing to speak of the greatness in Leonardo's painting of *The Last Supper* (figure 28) – a judgment I can only accept from those better qualified to make such pronouncements – and quite another to point out the devices which he used to isolate Christ at the moment when he declared that "One among you will betray me." The most obvious device, of course, is the placement of Christ at the vanishing point of the one-point perspective system used in the structure of this painting. In addition, Christ presents a dark silhouette against the light of the open window behind him. It also helps that the disciples immediately at his sides are leaning away from him. I do not know if these devices make this a great work of art or not; it does not really seem to matter. What does matter is that I can now see the painting with a greater understanding of the structure of the image before me and how the structure assists in creating the effect of isolation, of suddenly being alone among friends. This understanding by itself makes the painting more interesting and thus enriching even if I do not have an emotional or cultural involvement in its subject matter (which I do). It also provides me with some technical insights that I can draw upon when next I wish to try and work out an idea in a sketch or diagram of my own.

A much more explicit, but perhaps overdone, example of this can be found in a book by Watkins (1946), *Language of Design*. In this book, Watkins presents the novice artist with what he calls the "ABCs" of linear

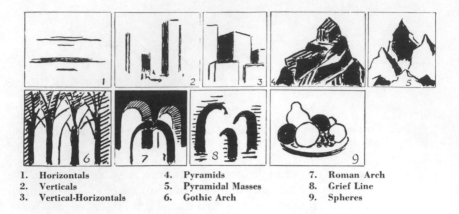

1. Horizontals   4. Pyramids   7. Roman Arch
2. Verticals   5. Pyramidal Masses   8. Grief Line
3. Vertical-Horizontals   6. Gothic Arch   9. Spheres

Figure 30 The "static motifs" of Watkins (1946,25). Courtesy of the Phillips Collection.
1 Horizontals: repose, calm, peace, or death.
2 Verticals: dignity, austerity, life, or power.
3 Vertical-Horizontals: stolidity, stubbornness, enduring.
4 Pyramids: enduring, majestic, mountain peak, keenness, alertness.
5 Pyramidal Masses (Pointed Shapes): alertness, penetration.
6 Gothic Arch: forest, contemplation, spiritual aspiration, mystery.
7 Roman Arch: enormous supporting power, weight, solidity, sombre.
8 Grief Line: fatigue, sorrow, grief, melancholy, bent with burden.
9 Spheres in Profusion: fruit, abundance, comfort, good life.

expression and nearly two dozen motifs with which one can express
graphically a great range of emotions and ideas. The six ABCs, shown in
figure 29, are ways in which line orientation or form can be used. Most
are as you would expect – a jagged line for excitement, a horizontal line
for calm, an unbalanced line for movement, and so on.

Based on these ABCs, Watkins describes nine "static motifs." Some are
but simple statements of the ABCs. Others are elaborations of individual
statements or combinations of them. The complete list with examples is
given in figure 30. He also describes four "motifs of movement": the
unsupported diagonal, italics, symbols for speed, and vibration; three
"expressions with curved lines"; and seven "dynamic motifs."

Taken all in all, the motifs represent at least one person's attempt to
codify some of the ways lines can be expressive. And while this range
of emotions is rarely expressed in conventional maps, the need for producing contrasts is very much a part of any graphic, whether it be in
pictures, advertisements, or the many kinds of maps that we can
identify and describe. In particular, the generalization of a line so that
it can, on the one hand, form part of an unobtrusive background of a
map or, at the other extreme, become part of the dynamic message of a
map can call upon many of these devices.

EDUCATION IN GRAPHIC
COMMUNICATION

Are there any other ways in which we can describe, in educational terms, the common ground between art and geography? Four curriculum goals have been identified by NAEA (n.d.):

1 *seeing and feeling* visual relationships;
2 *producing* works of art;
3 *knowing and understanding* about art objects; and
4 *evaluating* art products.

These goals are equivalent to perceiving, performing, appreciating, and criticizing. When viewed superficially, these goals seem much like those identified by Suzuki and Orff. In perceiving, we are dealing with a primary means of interacting with the world, in our case through vision. Information from and about the world is the basic ingredient from which maps are made; the processing of that information can be done either directly or indirectly. We do the former when we witness the world around us through our senses, primarily our visual system of eyes and brain. We do the latter when we process some representation of the world, whether that representation is numeric, analogic, or graphic. But even when we utilize some representation of the world, we process it through vision. Thus the eye-brain may be the most relevant system we have for discovering information about the world.

In recent years, we have become aware of the various ways in which the eye-brain gathers and processes information and of the tendency for these activities to be localized in one of the hemispheres of the brain. The localization may be different from one person to the next, but the occurrence of these activities can be associated symbolically with either the left or the right side of the brain. Edwards considers the activities in two categories – "Left-Mode" and "Right-Mode" as in figure 31 – and suggests their relationship to some educational aspects of hemisphere specialization and to the duality of *Yin* and *Yang* of Chinese Taoism.

Obviously all the ways of knowing listed here have a value and a place in human thought and consciousness. The argument for the maintenance of the arts (in general) in the curriculum is that they address more of the Right-Mode or right brain functions, whereas language and mathematics tend to address only the left. Even in engineering curricula, there has been a decline in the apparent regard for design courses, despite the historically positive role which graphics, that is to say, nonverbal thought, have played in the history of technology (Ferguson, 1977). In most eloquent statements, Edwards (1979), and

L-MODE	R-MODE
verbal	nonverbal
analytic	synthetic
symbolic	concrete
abstract	analogic
temporal	nontemporal
rational	nonrational
digital	spatial
logical	intuitive
linear	holistic

Figure 31 A comparison of Left-Mode and Right-Mode ways of knowing and processing information (Edwards,1979,40).

others,[1] are arguing that we are educating only one-half of the brain.

As with music, there is no question of whether children can produce things in art or, more significantly, do things that require them to look with discrimination and skill, just as they can listen in these ways. In fact, it is claimed (Mackworth and Bruner, 1970, 171) that children are smarter with pictures than with words. A goal of education, then, would be to build on and improve these skills for they contribute to skill development in other activities.

However, there does not appear to be an educational approach to art that systematically, and overtly, examines "looking" as an active perceptual process. But Mackworth and Bruner also point out that adults are smarter with words than pictures and wonder why this turnabout takes place. Two of their answers are of interest here. The first is that the slow initial learning of verbal material may arise from the fact that the repertoire of familiar visual images may be much greater than that for words, for reading is usually concerned with only a few thousand words. But in the long run, pictorial learning may lag behind verbal learning because pictures may have to be stored as double-entry items, both as words and images. Since visual input is always changing, it is probable that naming pictures helps recognition, but such naming requires a great deal of time.

Coulter (1985, 90) suggests that up to the second grade learning remains inextricably linked to movement and experience; the child must participate in any real learning. Obviously pictures and landscapes are easier to experience at that age than words. Another explanation for this differential learning may lie in the innate ability of children to process spatially arrayed information, that is, to deal with it in some holistic way, and that, somewhere along the educational line, they lose this ability. Edwards (1979, 62) contends that today's typical adult, one who has not studied drawing, draws at the nine- or ten- year-old level, like a child. Similiarly, Coulter (1985, 92), in describing the brain's developmental timetable, notes that for many adults the ages nine to eleven represent the "state of their arts" – the height of their development in art, music, athletics, grace of movement, or creative writing. She relates this situation to the completion of the corpus callosum which ends the easy access to the right hemispheric skills, which heretofore had been involved equally with the left hemisphere in most tasks.

Teaching beyond this crucial age must obviously have to compensate for this developmental factor; teaching in the years before must equally take advantage of this redundant power. In other words, education through the first five grades should promote the symmetrical growth and development of processes on both sides of the brain by allowing children sufficient time to examine various kinds of graphic images, to study the ways in which these images make use of variations in image content and structure, and how they convey information.

But there is an important distinction that is forced upon music by the fact that not all tone combinations are generally possible or particularly pleasing. As a result of this, there is formulated in our musical literature, and in the instruments themselves, a relatively select number of sounds that have, over recent centuries, been found useful and pleasing to composers and audiences. Not all cultures have agreed upon the same specific tones, but all have been selective (O'Brien, 1989). It is this essential restriction on creative options which would seem to set apart musical performance from production in the graphic arts. This is particularly true when the latter is set within the context of this statement of NAEA (n.d.):

One of the traditional and unique functions of the arts has been to emphasize individual interpretation and expression. The visual arts today continue to be a means whereby man attempts to give form to his ideas and feelings and to gain personal satisfaction through individual accomplishment. The growing complexity of our contemporary culture, including its visual aspects, also requires of every individual a capacity for visual discrimination and judgment.

Aspects of Creativity

Perhaps the definitional problem lies in the way we regard creativity. Certainly there is much confusion in the literature. On the one hand, we see an art teacher promoting art as illustration in geography (Lindsay,1975, 55-7), especially where the art student (and not the geography student?) is given ample scope to be original and to be creative. And in a most telling remark, Lindsay feels that an integration of art and geography in certain selected topics would allow the student to see "that the processes learned in geography and the technology learned in Art are valuable." On the other hand, we see a cartographer declaring (Collinson,1981) that a cartographer is only a cartographer when he is being creative, when he engages his emotions in his task, and when he no longer imposes his will on the work at hand but allows his inner vision to guide his working! One wonders if they are talking about the same kind of creativity, and what its definition might be.

If we are to believe Poincaré (Freeman,Butcher, and Christie, 1971,45), creativity depends upon a protracted period of diligent, conscious work; a receptivity to the subconscious ideas one's mind turns up; and the application of special techniques to those ideas. Creativity in the first two definitions above would not seem to involve much diligent work, such as the collection of data, the definition of the problem, and perhaps several attempts to reach a solution.

Perhaps then, there are two kinds of creativity that people think about. One is the complete freedom to think up something, to provide a product in the absence of any apparent constraints. This situation is, of course, extremely rare for we are always constrained in some way by time, space, or materials. On the other hand, if children are provided with working materials in an open-ended, unspecific task situation, they are being given essentially complete freedom. In this instance, children are constrained by the materials provided, but not in the way they were advised to put together or work with these materials. As a result, there is no inherent way to "know, understand, or evaluate" their finished products. This presents a real dilemma for today's art teacher who wants to play an active role in structuring the learning environment and yet wishes to respect the rights and freedom of the child which were "granted" in the Progressive Era of art education (Gaitskell and Hurwitz, 1970, 29).

There is, however, another type of creativity, where the materials are also constrained but the student is given a specific goal: How can you use these elements to produce not just any effect but this specific one? This is a type of creativity which closely resembles the process of improvization in music. It provides ways in which both students and teacher can know, understand, and evaluate a graphic product. The difficulty of this latter approach is that the teacher must be clear as to why those elements were chosen and what intellectual groundwork is being laid for subsequent lessons. As Suzuki stated, from the beginning, every step must be precisely identified and thoroughly mastered.

Figure 32 The intersecting processes of seeing, imagining, and drawing our world.
After McKim (1972,6).

Seeing, Imagining, and Drawing

The essence of all these activities in visual thinking – seeing, producing, knowing, and evaluating – has been captured best by McKim (1972) in the simple Venn diagram shown in figure 32. There are three complimentary activities in one's interaction with the world: (1) that between seeing and drawing, (2) that between seeing and imagining, and (3) that between imagining and drawing. The first, which we probably think of most often, involves skills in eye and hand coordination, the accuracy of looking, and, Edwards would say, in our ability to see with both sides of the brain. The second involves skills in the accuracy of looking and memory, and particularly accuracy in remembering aspects of the scene or object that are characteristic or significant of it. The third also involves eye and hand coordination, and the ability to obtain from memory the characteristic or significant attributes of the scene or object to be drawn. Carswell (1986) uses the more useful term "modelling" to include other forms of expression than simply drawing.

Thus training in the graphic arts should be concerned with all three interactions. Perception, in the sense of improving our perceptual abilities, would involve tuning the eye and improving its ability to look for and discover the subtleties of detail and structure. This is the ability that sets apart skilled observers from the rest of us. This is not necessarily training in acuity, although Seiderman and Schneider (1983) provide an excellent review of the application of eye training in athletics. One aspect of memory that is involved in these interactions is the intellectual connections that can be made between such things as the physiological, psychological, or conventional meanings that we can ascribe to various symbols or components of graphic images, namely, the physical, Gestalt, or cognitive properties of the symbols we use.

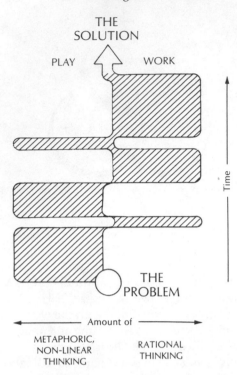

Figure 33 The two types of thinking involved in problem solving. After Samples (1976,76).

Another aspect of visual thinking is implicit in the diagram found in figure 33. Here Samples depicts problem solving as a combination of work and play, with most of the play coming at the beginning, most of the work coming near the end when the solution is in view. The words work and play are, of course, symbolic of several kinds of thinking and knowing.

Play can be interpreted as lateral thinking, right brain, holistic thinking, or doodling. Graphic designers do this all the time when they make a series of trial sketches or layouts before deciding on a solution to a graphic problem. Thus play is a random, searching process that calls upon all sorts of associations and relationships in the hope that some combination will suggest a new way to work toward a solution to a problem. Work, in contrast, is usually considered directed or systematic thinking, left brain, analogic thinking, or scaled drawing, as in architectural drawing or topographic mapping. Thus work is controlled, systematic effort channeled in a narrow way along some path toward some goal.

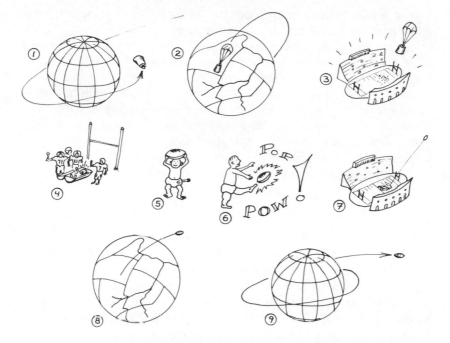

Figure 34 A sequence of nine narrative images (by the author) leading from a satellite circling the earth. The obvious influence of Superman is seen!

Obviously, there are graphic and conceptual components to both work and play that we can utilize in problem solving. Both approaches should be a part of our education in graphics and visual perception. Perhaps the most effective way of revealing these kinds of properties and associations is by giving children and students the opportunity to apply them in real situations with problems to solve or stories to tell. Ehrenzweig (1969, 45) considers that diagrammatic representation (as opposed to a precise, finished drawing) serves two quite different purposes. It provides an opportunity to set forth the essential elements of a finished product. But also it aids in the search of a not yet existing solution; it is a condition in which the design must remain vague and open-ended so as not to pre-empt a solution at too early a stage. Clearly, seen in this way, drawing becomes much more a form of language – a way of expressing and evaluating concepts and ideas (Litt,1977,75).

A well thought out and practical variation on this theme can be found in the narrative drawings described by Wilson and Wilson (1982). There are several versions of this activity. One involves the interaction of a student with his teacher or another student. The first person draws a simple scene or situation and then passes the paper to another person who responds by adding to the scene or by creating a second scene which carries forward the action of the first scene. The paper is passed back and forth, and each time the action advances in unpredictable ways. An example of one result of this activity, created by the author while in a workshop conducted by the Wilsons, is shown in figure 34.

In the cartographic area, similar activities are suggested in *Thinking About Ontario* (Castner et al., 1981, 14). Here map outlines are used as the background for recording movements of various people as a way of (1) telling a story that could then be interpreted and described in words or (2) illustrating a story that has already been written.

CONCLUSIONS

The need and value of classroom experiences in such things as visual perception, drawing, and visual thinking are not in question. How they can be introduced and maintained as a part of the curriculum is a problem to which educators in both art and geography must respond.

It would seem that at the level of school boards and funding debates, a curriculum designed to advance the child's ability to look critically and to be fluent graphically (to be graphicate) would be a most saleable commodity. This would be particularly so with those who are in a position to make significant cuts in educational budgets and who also consider art (and the arts) as a frill. It would be easy, and I would think beneficial, for art education to promote its role in developing perception and visual thinking.

Meanwhile, cartographers should certainly be interested in demonstrating the value of all kinds of maps for developing graphic (and geographic) thinking and problem solving. In addition, there are many areas in which we have a natural interest and in which we can promote the general skills and concepts in visual perception and graphic design. These include such topics as symbol design, transformations, perspective, color, and texture. Perhaps one day we will see these facets as important components of a new curriculum that acknowledges the value of visual discrimination skills and encourages the development of graphic thinking.

CHAPTER SIX

The Science
Connection

When I think of the formal scientific method an image sometimes comes to mind of an enormous juggernaut, a huge bulldozer – slow, tedious, lumbering, laborious, but invincible.   Robert Pirsig, (1981, 92)

One of the curious features of modern physics is that in spite of its overwhelming practical success in explaining a vast range of physical phenomena from quark to quasar, it fails to give us a single metaphor for how the universe actually works.   Nick Herbert (1987, xi)

The most productive form of learning is problem-solving. The traditional method of confronting the student, not with the problem but with the finished solution, means to deprive him of all the excitement, to shut off the creative impulse, to reduce the adventure of mankind to a dusty heap of theorems.   Arthur Koestler (1968)

Much of the creative thought of the designers of our technological world is non-verbal, not easily reducible to words; its language is an object or a picture or a visual image in the mind. It is out of this kind of thinking that the clock, printing press, and snowmobile have arisen.   Eugene Ferguson (1977, 835)

What is science? The definition that it is the search for explanation suggests that there are things we do not understand. But this is rather a negative way to consider all that we do know; after all, there have been some rather stunning scientific achievements. But there may be a certain amount of illusion about the apparently stable, smooth-running environment that man has constructed around himself. Perhaps the most exemplary expression of this illusion is in the public's attitude toward health and medicine. There are many individuals who seem to feel that they have no obligation or role to play in maintaining their bodies in good working order. They feel free to abuse themselves in the belief that medical practice can always be relied upon to make things right again. Fortunately, medicine's record here has been very good, especially in the broad areas of public health and disease control. However, one can read in the newspapers every day of individuals whose expectations of the medical enterprise were not fulfilled and who have sought recompense in the law. But there is something very positive about this faith in the omnipotence of science as we contemplate harnessing new forces, attractions, and reactions in nature, or constructing new buildings, bridges, and other engineering wonders.

But perhaps from an educational point of view we have come too far in tolerating this idea of omnipotence – an idea which can only reinforce Pirsig's description of formal science and lead to a "ho-hum" attitude toward that "bulldozer" when it is presented in the classroom. Perhaps then an awareness of the uncertainty of science is what is missing, and with it all the excitement of scientific research and discovery. This chapter examines some evidence of the nature of scientific knowledge and how it is taught. From such an examination, some implications can be drawn as to the positive relationships between geography and science, and at the same time how we might avoid in geographic education some of the problems that have become a part of science education.

THE NATURE OF SCIENCE

Jerome Bruner recounts the remark of James Conant, made at a Harvard faculty debate, that the object of science is to reduce empiricism (1962, 74). The scientist should be creating rational structures and general laws that, in the mathematical sense, predict the observations we would be forced to make if we had no general laws. As a result, science increases the unity of our experience of nature and "is the hallmark of the way of knowing called science."

Intuitively there is a conflict here if one thinks of the "rationality of science" as opposed to the "metaphoric non-rationality of art." But in the previous chapter I tried to establish the value of the notion of drawing as thinking. In this context, and with Bruner's remarks in mind, science and art would seem to go hand in hand. This interactive operation would appear to function much like the process described by Neisser in figure 35. Here the general laws act like the mental schemata which help direct our perceptual exploration of the environment, which results in sampling the present and discoverable evidence, and which subsequently modifies the schemas which first directed our attention toward the environment. Bertrand Russell put the connection

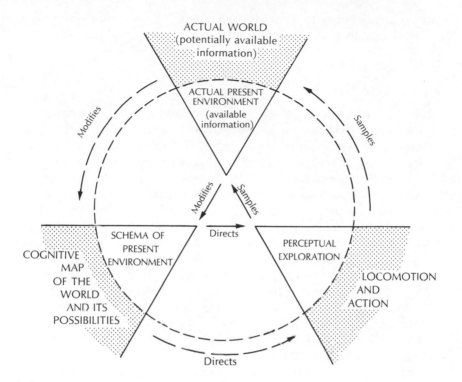

Figure 35 The triad of the mental structures, locomotion and action, and the actual world as they interact in a continuous cycle. In it, cognitive schemata 1) direct perceptual explorations which 2) sample the present environment from which information 3) modifies the original schemata which direct... . After Neisser (1976,112).

even more strongly when he said: "Physics is mathematical not because we know so much about the physical world, but because we know so little: it is only its mathematical properties that we can discover" (Bruner, 1962, 66).

Until we discover more, we must proceed by intuition and metaphor. We are also forced to do this because of the tremendous discrepancy between what we can know and what there is to know. Thus in order to produce a more efficient, economic, or useful expression of knowledge, we must perceive things systematically (rather than in detail), as classes (rather than individually), or as metaphors (the asymptote in mathematics can be depicted by Sisyphus forever pushing his rock to the brink of the incline). In this context then, it would be difficult to proscribe the processes of symbolization, classification, or model building exclusively to art or science. What we can do is to make use of the

much more systematic approach to knowledge acquisition and problem solving that is encoded in scientific thinking and the scientific
method.

The Scientific Method

Robert Pirsig (1981, 92) gives a clear and a useful view of science and the
scientific method in the context of motorcycle repair. He first defines the
difference between inductive and deductive logic. The observation that
the cycle engine misfires every time it goes over a bump leads to the
logical inductive conclusion that the misfiring is caused by the bumps;
one goes from particular observations to general conclusions. In contrast, knowing that the horn derives its power exclusively from the
battery, the mechanic can logically deduce that if the battery is dead, the
horn will not work. A deduction, then, starts with general knowledge
to predict a specific observation. The scientific method, in the case of
motor cycle repair (or any problem solving), is simply the correct
program of interweaving these two kinds of logic, utilizing (1) observations of the machine, and (2) mental structures of it found in the
manuals. Or, perhaps as I suggested in the previous chapter, it is an
interweaving of work (rational, linear thinking) and play (lateral or
metaphoric, nonlinear thinking).

For simple maintenance problems, a formal procedure is rarely
invoked. But for really complex or difficult ones, like those we set for
ourselves in trying to understand the world, a more formal procedure
is required, for after all; "the real purpose of scientific method is to make
sure Nature hasn't misled you into thinking you know something you
don't actually know" (Pirsig, 1981, 93).

The invocation of the scientific method, according to Pirsig, involves
writing down everything you know in a notebook under six categories:
(1) statements of the problem, (2) hypotheses as to the cause of the
problem, (3) experiments designed to test each hypothesis, (4) predicted results of the experiments, (5) observed results of the experiments, and (6) conclusions from the results of the experiments (Pirsig,
1981, 93). If you do not commit yourself to paper, you can get careless
or go romanticizing about the problem and Nature will soon make a
fool out of you.

Obviously, surrounding our observational activities with rules to
which we adhere is one way to insure we do not get careless. This is an
essential aspect of thinking, but, as Munby (1982, 12) points out,
adhering too zealously to rules can create other problems as well. In any
kind of problem solving or creativity, the trick is also knowing when to
break the rules. This particular "thinking skill" is very difficult to teach.

SCIENCE EDUCATION

What then are the goals of science education and the unique contributions which it can make? The Science Council of Canada (Science 36, 13) expresses the need for a science education in Canada that can:

• develop citizens able to participate fully in the political and social choices facing a technological society;
• train those with a special interest in science and technology fields for further study;
• provide an appropriate preparation for the modern work world;
• stimulate intellectual and moral growth to help students develop into rational, autonomous individuals.

In a review of Yearbooks of the National Society for the Study of Education in 1932, 1947, and 1960, Munby and Russell (1983, 160) noted three major themes in arguments in support of the science curriculum:

• students should know something of the body of scientific knowledge;
• they should develop skills of problem solving; and
• they should develop some of the attitudes attributed to scientists and to the climate in which science may continue to prosper.

These kinds of goals and themes suggest that teaching a scientific *method* may be more important than teaching the *content* of science, for there is nothing in these goals and themes that suggests the greater value of one component of study over another, as, for example, the greater value of Boyle's Law in thermodynamics or the right-hand rule in electricity. In fact, in their study, Munby and Russell refused to be drawn into any such designation of specific content even though the volume for which they were writing specifically asked them to focus on the question of a common curriculum in the natural sciences. Rather, they defined science (1983, 169) as a unique method "of using language to construct a highly organized and powerful way of looking at the world and understanding it.

As an example of this definition, Munby provides a number of classroom dialogues which illustrate something of the basis of scientific knowledge and the use of language. Part of one of these dialogues, from a grade 4 science lesson, begins where a teacher in introducing a classification system, has arrived at a distinction between living and nonliving things (Munby, 1982, 19-21).

Teacher: "Now, we are going to leave the nonliving things for later and study just the living things." (Writes "living" on the board.) "Now let's divide all the living things into two divisions. Into what two divisions can we divide every living thing? Every living thing is either a ...or a ...? Lucy, give me one division."

Lucy: "People?"

Teacher: "People are just part of one of the two divisions."

Peter: "Plants and animals."

Teacher: "Good for you, Peter. That's right. Every living thing in this world is either plant or animal. People, Lucy, are animals, so they fit in this division."

Lucy: "People aren't animals, they're humans."

Teacher: "People are animals, the same as dogs and cats and so on."

The dialogue continues with both the teacher and Lucy adamant in their contentions about the placement of people in the scheme of things. Others offer tidbits (people talk, cats do not), and apparently there is much discomfort over the teacher's hard line. Eventually the teacher closes the discussion with the declaration, "That's enough. People are animals," and continues on with the lesson. From the wreckage of this "lesson," Munby points out that neither party apparently understood that (1) the concept animal is just that, a concept, and (2) "animal" is part of two different classification systems or taxonomies – a scientific one and an "everyday" one. Having been constructed for different purposes, the taxonomies may give different meanings to specific words, like "animal." It is no wonder that Lucy and the teacher could not accept each other's logic because each was operating in another classification system which rendered the other's definition wrong, according to their own system. But of course they were both right within their own classifications.

If the teacher appreciated the fundamental idea that science is a way of creating realities and knowledge, then the problem would never have arisen. However, the teacher was in trouble much earlier in the lesson when she rejected Lucy's answer offhand. At the very least she could have accepted it *pro tem* , and the problems with it being one of the two major categories could have been discussed. It is obvious from the discussion that the class was able to supply other information about people and cats. In not too much time it would seem that the need for three categories might have emerged, namely, people, dogs and cats and so on, and plants. It is with this revelation that the truth of Lucy's and the teacher's two taxonomies would have become apparent. One might actually question whether, in the opening statement, the teacher should have been as specific as identifying two divisions. Why two? It seems a classic case of a teacher, and unfortunately it is not exclusive to science teachers, who knows what information has to be "taught" and is unable or unwilling to allow the children to "learn" a more fundamental concept instead or as well. When Peter supplied the "correct" answer, any learning probably ceased because it was now common knowledge what the teacher "wanted." For real learning to have taken place, it would have been better for Peter to have remained silent.

It is also of particular interest and surprise to read the words of those scientists who point out that science is *a* way, not *the* way, for other areas of human inquiry also construct reality in different ways, with different products, and with different ends in mind. That, they conclude, is why we have separate disciplines. After all, science is not unique among the disciplines in using rational argument and critical thinking (Science 36, 17). Thus science cannot and should not be an exclusive vehicle for the development of these skills, especially as they relate to the study of the natural environment.

If we are to consider changes in the way in which we approach the teaching of cartography and geography, and we acknowledge our common interest with science in the natural environment, is there anything we can learn from science education experiences? Is there anything we can do to avoid the "juggernaut" image that Pirsig has envisioned?

Experiences in Science Education

During the late 1970s, a comprehensive study was undertaken by the Science Council of Canada (Science 36; Science 52), in order to establish a documented basis for describing the present purposes and characteristics of science education in Canada. How typical or applicable the findings are to the United States is difficult to determine precisely. But in accepting their general truth (and I do not think that is asking very much), they provide considerable insight into the kinds of problems faced by science education. Four critical areas were examined (Science 52): overall problems, textbooks, how science is actually taught, and issues for deliberation.

The first of these areas, overall problems, concluded that science "was being taught as a body of knowledge and technique, without any mention of its personal, social or national relevance," as though it were "all- and self-sufficient." Concern was expressed that schools were perpetuating the separation of arts and science by ignoring the relationships of science to other curriculum subjects in specific, and that of science, technology, and society in general. The reports chide current teaching practice for not producing "independent thinkers who understand properly the basis of their knowledge," and who have a "skeptical, divergent, questioning, and imaginative approach towards the solution of problems."

The second area, textbooks, concludes that schools teach that scientific inquiry is basically inductive in nature and is based on cooperation and communication. Yet laboratory sessions are not organized to favour this approach. While most textbooks properly describe the scientific method as including the definition of the problem, observation, gathering of information, formulation of hypothesis, designing the experiment with controlled variables, verification, and communication of results, "ninety-five per cent of experiments suggested in the textbooks are highly structured. Students are seldom asked to formulate a question or define a problem. Laboratory manuals used in senior grades generally ask students to verify laws previously learned in class (the deductive approach) rather than to generalize from information they themselves have collected (the inductive approach)." Munby and Russell (1983, 160) make the more disturbing comment that: "What is taught in schools is closely related to what is written in textbooks, but the basic characteristics of our textbooks do not change. Content has been updated to keep pace, to some extent, with the development of new knowledge in the sciences, but textbooks are at their best in presenting statements of fact and interpretation which are to be learned by students for later use in responding to teachers' questions, both verbal and written."

In examining the third area, how science is actually taught, the report notes that there is an emphasis in the middle years on "covering a considerable body of material" – this expression "means that the 'correct' explanation must be included in students' notes." For teachers at this level, accuracy is at the heart of what they believe to be a scientific approach to problems. Thus the emphasis on approved explanations and the right answers is at odds with both the process of inquiry and the conceptual and tentative status of knowledge in science.

Perhaps the most sympathetic statement about the real world of the science teacher, if not all teachers today, is contained in the following: "Confronted with uncertainties about subject matter, student behavior and educational goals, teachers approach their work in ways that will make it less uncertain, thus accommodating to complex situations over which they have no control" (Science 52,24).

Finally, in addressing the fourth area, issues for deliberation, the report asks whether the absence in guidelines of a stipulated hierarchy of process skills to be taught leads to a danger that only the lowest-level skills will be attended to? It also wonders how teachers can integrate the subject matter of science with that of social studies, mathematics, and the technical fields.

Pirsig also gives us, in the quote at the outset of this chapter, some insight into what may be an underlying problem with science education. He completes his image with the thought that it may take two, five, or a dozen times longer to solve a problem using the traditional scientific method than it does by employing the informal techniques of a mechanic, but you know that in the end you're going to get it (1981,92)! He also notes that the care necessary in formulating these various statements often results in much scientific writing appearing to be dull and, he even uses the word, "dumb". Perhaps this is why one educator finds the scientific method so "tiresome," and that in the history of science it does not seem to have led to "mind-boggling conceptual novelties" (Munby, 1982, 30). One only has to read Burke's fascinating book *Connections* to find a wealth of examples of the apparently illogical jumps which creative thought has taken in the name of scientific creation. For example, Burke describes (1978,72-5) the convoluted sequence of inventions and adaptations that started with the problem of flooding in deep shaft mines in Wales. Eventually, its solution, with the perfection of a suitable pump, led to the invention in Italy of the barometer, which in turn multiplied the number of possible routes that further scientific innovation could take.

Pirsig is particularly concerned with step three of the scientific method, the experiments, which are often seen in isolation from the rest of the steps. As a result, the primary impressions made are of test tubes, bizarre equipment, and the people "running around making discoveries." He goes on to make the rather sharp comment: "A man conducting a gee-whiz science show with fifty thousand dollars' worth of Frankenstein equipment is not doing anything scientific if he knows beforehand what the results of his efforts are going to be. A motorcycle mechanic, on the other hand, who honks the horn to see if the battery works is informally conducting a true scientific experiment. He is testing a hypothesis by putting the question to nature" (Pirsig, 1981, 93).

School Science

Another aspect of the educational problem is raised in *Mind-storms*, Seymour Papert's fascinating little book on education and the computer program, Logo. In this book, he makes the distinction between "mathematics," that vast domain of inquiry whose beauty is rarely suspected by most nonmathematicians, and "school math," the product of historical accidents which have determined the choice of topics that should make up the mathematical baggage citizens carry around (Papert, 1980, 64). He describes school math as a *QWERTY* phenomenon – something that has evolved in part from the tendency for the first useable but still primitive product of a new technology to dig itself in and become the convention. *QWERTY*, of course, refers to the first five letters at the top left of the standard typewriter keyboard.

The Science Council of Canada's Report (Science 52) also mentions "textbook science," as though the problem is not exclusive to mathematics. The three reports we have been reviewing suggest that the scientific community has already tried to determine just what constitutes "school science" and to answer such questions as: "How many present-day scientific conventions are *QWERTY* phenomenon?"; "If there are any, how did they come to be?" In other words, what "scientific baggage" do we want our citizenry to carry around?

NATURE OF SCIENTIFIC KNOWLEDGE

What then can we regard as a positive educational way of viewing scientific knowledge? Munby reduces the difficulty of answering this question by pointing out that all disciplines construct their own realities out of their unique networks of concepts, principles, explanations, and theories, all of which are human inventions. Each is, in its own way, rule bound.

Science is particularly interested in constructing realities (generalized models) of natural phenomena so that they can be discussed, explained, and their behaviour predicted. To do this, the constructs should match the data consistently and be free from the effects of "willful behaviour" (Munby, 1982, 22). Frequently, scientists must use mathematics to state their constructs as precisely as possible. Yet these constructs remain inventions of the human mind. As such, they are not, strictly speaking, true or false, right or wrong; rather they conform or fail to conform, to our observations and thus are judged on the basis of their usefulness in our attempts to construct realities that match our observations. As a result, there will be many theories that are of variable value and many which will fall into disuse. Such is the nature of our attempts to explain and generalize our observations about the world.

IMPLICATIONS FOR GEOGRAPHIC EDUCATION

I have examined some serious problems which have been raised about science education. Certainly we should try to avoid repeating them in education in cartography and geography. On the other hand, the view of science as an intellectual activity aimed at constructing generalized models of the world seems to provide a way around many of these problems.

Stated as a question: Should we be teaching the constructive processes of scientific modelling or the facts accumulated by those processes through history? The same question might also be raised about the teaching of mathematics. The advantage of teaching the processes, both inductive and deductive, is that it forces one into a context, that is to say, into a time and place where a specific question is being asked within constraints that are probably specific to that time and place. Thus the study of that particular problem, for example, in its historical context, brings it much closer to our concept of improvization – of working out a solution within the context of particular constraints. In this context it would be useful to examine the other solutions that are possible or probable. Without this contextual setting, the "facts" emerging from the scientific inquiry will lose much of their meaning or may be misunderstood if they are interpreted in light of our contemporary knowledge. Certainly many scientific questions raised in the past appear silly when seen in light of what we now know, that is, out of our present context. Similarly, the political success of Senator Proxmire's "Golden Fleece Awards" depended upon the silliness of seeing serious research titles taken out of context. But the questions asked historically were not trivial and thus are still valuable from an educational standpoint. In addition, their very reality makes them of interest and a significant part of the history of how we came to think of the world as we do now.

Geography, if we consider it as an integrated discipline, has by its very nature the ability to examine scientific problems in contexts which provide the kinds of broad environmental or societal relevance that much teaching in science seems to be lacking. In addition, if this integration includes both art and science, then geography may be one of the few areas in which their interaction can be experienced and studied. On the other hand, there is much evidence that geography has lost sight of its integrating principles. We see the proliferation of subgroups and subdivisions of the discipline. The Association of American Geographers now lists nearly forty interest groups. As a result, more and more geographers can cater to and be labeled by their specific interests. It may be then that we are focusing increasingly on the "facts" and specialized applications and expressions of the integrating principles as they appear in our subfields.Consequently, we are in danger of becoming a microcosm of all intellectual endeavor in both positive and negative ways. On the one hand, we are collectively familiar with and have access to the ways most other disciplines construct their models of reality, albeit ours is a special interest in aspects of their spatial expression or operation. But on the other hand, in doing this we may be building into geography the kinds of intellectual blinkers which impede the communication among many disciplines. In so doing, the similarities and differences among their constructs of reality are more difficult to see and appreciate.

As noted above, scientists have acknowledged that the application of problem-solving and critical thinking skills are not unique to scientific education. In their most thought-provoking discussion of the common curriculum and a response to individual differences among students, Munby and Russell (1983, 174-5) argue that if teaching is.

to be conducted in a manner that is congruent with our aspirations for rational, moral and authentic encounters, then it must make provision for students to make their judgements from a position of knowledge. In its ideal form this requires that classroom discourse contain arguments, evidence, alternatives, and so forth, of the sort which we normally associate with rational discourse, and that these requirements become the responsibility of the students as well as the teacher.

Without alternatives, the experience loses some of its integrity and tends toward the inauthentic. This seems but another way of expressing some of the same ideas that were described in chapter 4. Improvization, in the music education realm, was a way of examining alternative solutions to a problem in such a manner that the student was party to their evaluation. In Orff, of course, there was the added dimension that the student was also involved in the generation of the "question" and the production of the alternative solutions.

Certainly a process approach to geography would provide, particularly in the early school grades, a way of addressing the unifying concepts of geography without denying the systematic sides of it. These concepts could be illustrated by the choice of content, that is in the questions raised, and in the contexts in which they reside. The idea of a process approach is not new (see Satterly, 1973, 168), but the recent research experiences in cartographic communication may provide the useful basis on which such an approach to the integrating principles of geographic thinking might best be taken. A consideration of this possibility is the subject of the next chapter.

CHAPTER SEVEN

Influences of Cartographic Communication

The majority of the atlases go to varying lengths in the first few pages to explain scale, distance, and direction, but in most cases guidance is severely limited by a lack of explanatory text and a somewhat mysterious collection of illustrations. It then appears to be accepted that once these obstacles have been overcome the atlas becomes entirely within the comprehension of the child.

Patrick Sorrell (1974, 84)

The process of map reading, like the process of reading text, requires the existence of structures in the reader's mind that are, at the least, equally as important as the marks on the paper. These mental structures vary with the map use task, but are identifiable and are finite in number.   Grant Head (1984, 2)

The growth of cartography as an academic discipline and as a research activity has been particularly vigorous during the last quarter of a century. Some of the factors contributing to this have been reported elsewhere by Castner (1983b), Muehrcke (1981), Petchenik (1983), and Wolter (1975). Suffice it to say here that this growth was driven in part by a desire to better understand how maps interact with the people who use them. In other cases it may have been the desire to discover and confirm the science behind the art in cartography.

In any case, there is emerging a body of theory and research in what is called cartographic communication. The visible products of this work include a variety of models of cartographic communication, a descriptive vocabulary of the component processes of that mode of communication, a variety of workable research methodologies, and a number of general guidelines for the purposeful use of various graphic elements in map design. The invisible products of this work include a greater awareness and respect for the role of the map user in the communication process, and perhaps a certain amount of humility

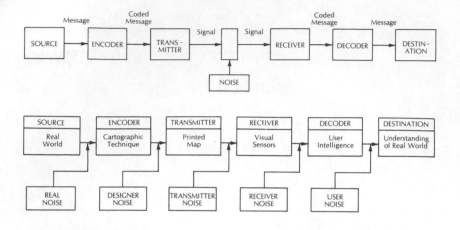

Figure 36 Two generalized communication systems in terms of (a) information theory (above) and (b) its cartographic translation (below). After Jolliffe (1974,176).

about the limits within which we, as map designers, can influence the flow of information from our products.

This chapter presents a selective review of these models of cartographic communication and a more thorough discussion of the various processes which describe map communication. This, in turn, leads to considerations of the implications for (1) cartographic design practice and (2) cartographic education. This latter consideration suggests the close relationship between mapping and geographic thinking.

MODELS OF CARTOGRAPHIC COMMUNICATION

The idea of diagramming the cartographic communication process came to us from studies in electrical engineering and information theory. The traditional model, figure 36a, involves encoding a message and transmitting it by a transmitter. The transmitted signal is subject to noise or distortion before it is picked up by a receiver, where it is decoded and presented as a message at a destination. It soon became evident, however, that such a diagram had to be modified, as in figure 36b, in order to account for all the sources of noise in a cartographic communication system and to capture the essence of the representational nature of mapped information. As a result, a great number of diagrams have appeared attempting to elaborate or illustrate various aspects of the cartographic communication process. In all, I have encountered more than a dozen such diagrams.

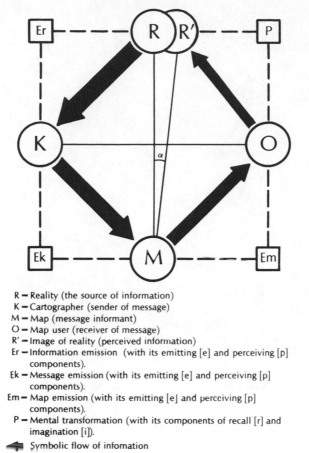

R — Reality (the source of information)
K — Cartographer (sender of message)
M — Map (message informant)
O — Map user (receiver of message)
R' — Image of reality (perceived information)
Er — Information emission (with its emitting [e] and perceiving [p]
    components).
Ek — Message emission (with its emitting [e] and perceiving [p]
    components).
Em — Map emission (with its emitting [e] and perceiving [p]
    components).
 P — Mental transformation (with its components of recall [r] and
    imagination [i]).
◄■ Symbolic flow of infomation
 α Degree of information loss

Figure 37 A model of cartographic communication. After Ratajski (1973, 219).

For instance, figure 37 is an example of a model in a circular form
which identifies 1) the processes of communication, 2) the players, that
is, the people involved, and 3) the products, namely, the forms of information. All are sources of noise. The circular form with the partially
overlapping circles, R and R', suggests the desire that the map be able
to produce in its user's mind a reasonable facsimile of the reality that
was used to produce the map. The degree of overlap would provide a
measure of the efficiency of the communication, as in the relationship
between M1 and M2 which was discussed in chapter 5.

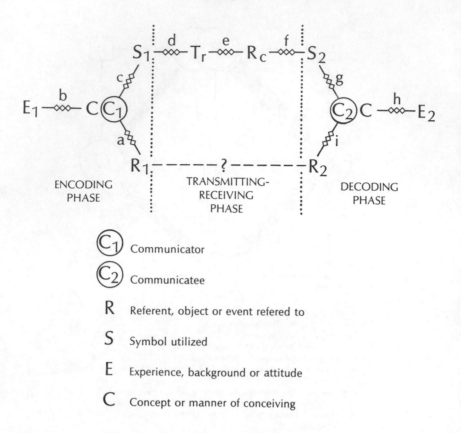

ENCODING
PHASE

TRANSMITTING-
RECEIVING
PHASE

DECODING
PHASE

$\widehat{C_1}$ Communicator

$\widehat{C_2}$ Communicatee

R   Referent, object or event refered to

S   Symbol utilized

E   Experience, background or attitude

C   Concept or manner of conceiving

Figure 38 A general communication model, modified from Alexander (1967,16).

The origins of such a diagram can be found in Alexander, figure 38. Here, the distinct phases of the information theory model have been broken down into the component sources of noise or error which are indicated by the "figured linkages" between each component. Alexander describes each in some detail in the general context of language and thinking. His diagram is a mirror image, top to bottom, of Ratajski's model, but without the overlapping circles for R1 and R2. Both have evolved into the more simplistic and more easily described model in figure 39, which "reads" something like this. The world, or reality, is sampled for information about some particular topic. This data set is usually gathered and processed by a map author in order to make a particular generalization or show a particular picture. The map maker translates that data into suitable graphic elements in a map. The map, whether in printed or electronic form, is viewed by a map reader who constructs his own image of reality from that viewing. It is our hope that

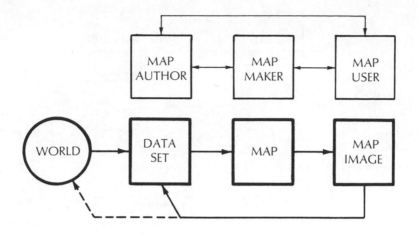

Figure 39 A simple model of cartographic communication which identifies the domains of and necessary feedback among map author, map maker, and map user(s).

the map reader's image of reality bears a close resemblance to that of the data set, and thus, by extension, to the real world beyond. In order for this communication to be most effective, it is necessary that there is a certain amount of communication or feedback between the map author, map maker, and map users.

Other models have appeared which are simply variations on these basic forms and ideas. The one exception is the model in figure 40. It represents an attempt to describe the map design process in terms of the economic realities of the cartographic workplace as well as to show the relationship to design of both theoretical and applied research. The former is represented by research in map perception and cartographic communication. The latter is represented by research in the various technological sciences that apply to the data gathering, data processing, and map production processes; areas such as remote sensing, computing, typography, and photography. Figure 40 shows (1) the initiation of a specific map request, (2) the design proposal and budget resolution loop leading to the acceptance of a design and the initiation of its production, (3) the input of perceptual research into design considerations, (4) the input of technical research into production considerations, and (5) the role of user feedback in the design process.

The value of such diagrams, however, has not been in the precision with which any one of them has been able to describe the communication process. Undoubtedly, that has been the goal of a number of such efforts. However, I believe their collective value has been in shedding

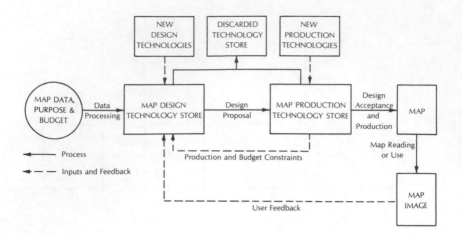

Figure 40 A model of cartographic communication devised by Eastman (unpublished) from a model by DeLucia.

light on the numerous ways in which communication can be considered. Specifically, the information theory model accounted for noise in only one place in the system. But, we have come to recognize that in cartographic communication there are numerous places where "noise" can disrupt or alter the process. Each model that I have seen seems to shed light better on some aspects of the system than on others. Thus in some discussions one is more useful than another. But if taken collectively, they represent both the vitality of our search for understanding, and the breadth of interest that the concept of cartographic communication has generated.

Given such variety, it is obvious that there is much more to the problem than we had allowed ourselves to consider. Of greater importance has been the realization of how much the map user contributes to the success or failure of the communication process. One of the benefits of knowing about our bipartite visual system, described in chapter 3, is that its structure can be applied to strategies in map design, particularly to creating appropriate hierarchies among the categories of mapped information.

In chapter 3, the concepts of pure and cognitive perception were described. From these concepts and other evidence presented there, we can envision a two-part visual system which deals essentially with sets of quite different visual tasks. On the one hand, foveal vision deals best with the inspection of the finer details and with the precise identification and verification of information. This might involve any scrutiny which requires matching some list of prescribed characteristics, dimensions, or conditions that are required for an identification, and simple counting or estimating procedures. All these activities involve a certain amount of conscious attention. In contrast, the global aspects of a visual display or image are best processed in peripheral vision. There, small stimulus points, lines, and areas of relative homogeneous tones or textures are assessed for their potential in yielding information relative to a problem at hand (for gathering specific pieces of information). Promising locations may then be more closely inspected in foveal vision. Most of this global processing is performed with little or no conscious effort on the part of the observer. Of course, the two "systems" work hand-in-hand and are undifferentiated in our consciousness unless we purposely attend to the distinction.

In map design, we can take advantage of this distinction by evaluating the role of each class of information that we propose to include and "assign" its processing to one of these two systems. For this, the concepts of map or subject information and base information are useful. To understand them, we must first address the questions: (1) What information will be required for a given map?; and (2) What specifically will we want the users to do with it?

ACTIVITIES IN MAP USE

For any particular map, we should first determine all of those things which the map user might be expected to want to do with it. We can call these the *activities* in map use. They are important for they speak directly to the information content requirements of the map. With the activities in mind, we can identify what information is central to their execution, what information is supportive of them, and what information is irrelevant. Knowing this, we have a basis for assigning graphic importance to each piece of information that we retain and, in the case of the irrelevant information, for removing it from the map altogether.

Establishing the map use activities is a science in itself. A considerable amount of research in map design and map use focuses on these particular questions. Such research usually features formal surveys that attempt to define the map use activities with clarity and precision and in a systematic and rigorous manner. In the case of series, special purpose, or reference maps, such information can probably be determined only through close contact with experienced and map-conscious representatives of the user groups involved. For most thematic maps, the map author is probably the best source of this information. In both cases, it is useful to incorporate in one's questioning demonstrations of the activities which the map is expected to support. For complex maps, a penetrating analysis of the elements of map, border, and marginal information must be considered as essential aspects in the performance of each activity (Castner and McGrath,1984). Only when all this has been accomplished can there be a definitive statement of the content of the proposed map.

It is at this point that a further distinction can be made among the types of information destined for the map to clarify their relative importance to the proposed map activities. This distinction can then be related in the design process to the perceptual phenomenon of "figure-ground" and the design concepts of map information and base information.

Subject and Base Information

In vision, those elements in a scene, picture, or map which stand out and attract our initial attention are deemed "figure"; the background or field on which these figures are seen is called the "ground." In map design, we wish the symbols or elements representing the most important information, what we can call the map or subject information, to be seen as figure. In contrast, we hope that the symbols and elements representing the supporting and less important information, what we can call base information, are seen as ground. In other words, in the design process, we assign intellectual importance to the various pieces of information to be included in the map. Subsequently, we assign appropriate graphic importance to those elements used to represent the chosen pieces of information. We do this so that we can also take advantage of the two ways in which we process visual images.

Information in the highest category, the subject information, must be made most prominent so as to be easily isolated by the map user and given the detailed inspection necessary to clarify its exact nature and identity. Supporting information, on the other hand, should be symbolized in such a way that its presence can be detected without the need of direct foveal inspection. This base information can be more subdued so that it can be detected and discriminated largely in peripheral vision and not interfere with scrutiny of the more important subject information. The information supplied in this way is supportive in that it may provide information on place or location of the map, on other less important features or dimensions of the subject information, or incidental material. The graphic image presented by the symbolized base information provides, to a great degree, a visual structure or backdrop against which the entire map is searched and eye positions are referenced.

Without this division of responsibility, as it were, the eye would have

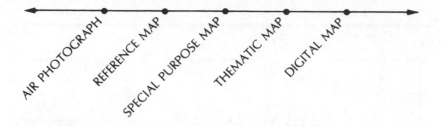

Figure 41 A continuum of map types based on the ratio of subject information to base information. The ratio increases to the right across the diagram.

to be far more systematic in its inspection of an image in order to determine priorities for closer scrutiny. This would be most inefficient and ineffective. We could liken this to the situation facing a blind person who must "read" a tactual map through his fingers, a task which requires processing the map point by point before he can begin to build up a total map picture from the many individual details. Reference back to particular places requires searching for the point, or leaving a finger of one hand in one place while the other hand goes off to examine other details. Obviously the power of our visual system is such that we can always start with the whole and then proceed to the specific, and for the most part this can take place without much conscious attention on our part.

A Design Continuum.   This distinction between subject and base information can also be used as the basis for defining, in design terms, various kinds of maps and graphic images commonly used by geographers. This is done by determining the relative amounts of these two kinds of information contained in the image and the effect of any modifying elements such as a title, legend, or associated insets. The continuum ranges from what we might call a digital map at one end, to an aerial photograph at the other, as in figure 41.

Consider first the category of reference maps and the most exemplary example, the topographic map, as in figure 42. Because of the wide range of possible users and potential uses of these maps, all of the information which they contain is nominally of equal importance. Thus in theory, all elements should be equally attractive or attention getting, and thus equally accessible to the user. In practice, however, we do make some distinctions stronger, as that between land and water and between certain features of the physical and cultural landscapes. Otherwise, all the other information is more or less equally important and should thus be equally accessible. The titles of such maps carry locational information, and their legends assist in the identification of

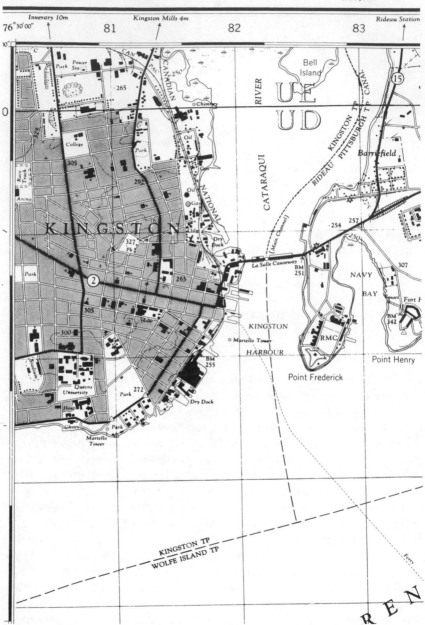

Figure 42 A black and white copy of a portion of a standard topographic map. This map is based on information taken from NTS map sheet 31C/1e ©1972 Her Majesty the Queen in Right of Canada with the permission of Energy, Mines and Resources Canada.

specific symbols, often conventional ones, found on the map. Technically, such maps contain essentially no subject information, only base information; as a result, the ratio of subject to base information is very low.

Other similar maps, in terms of this ratio, would be small-scale reference maps in atlases, city plans, cadastral maps, and the like. In the design of reference maps then, the stress is upon achieving maximum legibility without creating many strong figural images. Other ways of thinking of this is to say that there will be few hierarchies of information in the image, or that the elements are symbolized in such a way to allow the user to either "isolate" a class of information for inspection, or to ignore it when examining other information.

Near the other end of the continuum in figure 41, one would find typical thematic maps, such as the example in figure 43. The emphasis of this type of map is on a single message – to convey some single important idea, distribution, or relationship. The subject information must thus be made graphically prominent so as to produce a strong figural image. Usually the information is scaled in such a way as to promote only a few specific map use activities, such as simple ratio estimates, or to spot the location of particular subclasses. The supporting base information would be reduced in its amount, complexity, and prominence. As a result, the ratio of subject information to base information is relatively high. In the case of figure 43, regions of four socioeconomic levels are identified within a highly generalized structure of major roads and the coastline of the lake. The latter information provides a minimal locational framework for the more important regional information.

In the design of thematic maps, the reduction in the amount and complexity of base information is achieved by the selection and simplification of detail through generalization; its prominence is diminished by the reduction of its component symbols' contrasts with their environment by decreasing their size, boldness, or brightness. Reductions in prominence can also be achieved by changing the form of the symbolization so that it no longer requires foveal inspection to be detected and identified. For example, a linear grid, such as the graticule, will be much less conspicuous if it is reversed out of a tonal background (see figure 47 and its discussion for an example of this effect), or, terrain information can be symbolized by hill shading rather than contouring.

The titles and legends of thematic maps also provide some design opportunities to influence the communication process, opportunities that are not common to reference or special purpose maps. Titles and legends can identify, for example, the area mapped, the nature of the

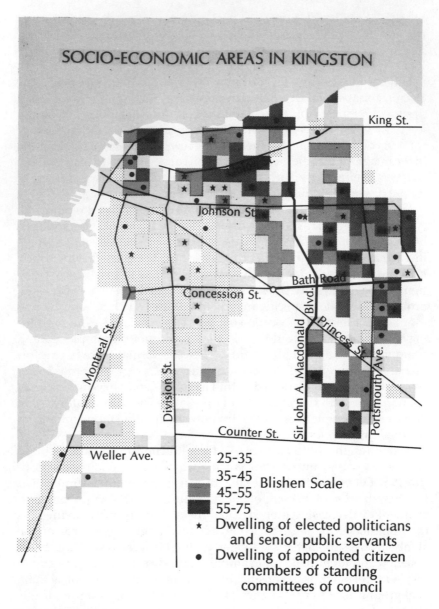

Figure 43 A typical thematic map.

subject information, and the map use activities implied. The titles of thematic maps are required to identify the nature of the subject information and perhaps the area mapped. I say "perhaps" because often the context of the map is established in other ways. For example, in a

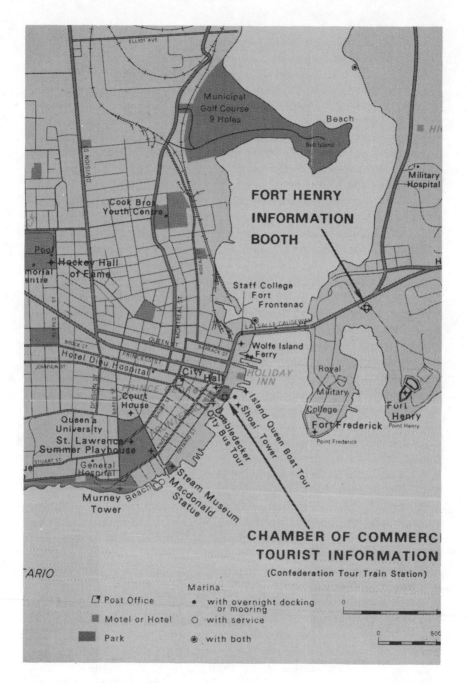

Figure 44 A portion of a typical special purpose map; in this example, a tourist map of Kingston, Ontario.

regional atlas every map need not mention the place name in its title. Or the familiarity of certain country shapes, for example, will convey sufficient locational information without having to specify it in words. This leaves the title available as a place where information about the map data and map activity can be suggested. For example, the use of the words "larger and smaller" would suggest that one should make ordinal comparisons with the map information. The use of the words "the dominant position of" might suggest that one class of data stands in isolation and magnitude above all other data classes. Obviously the mapped information should convey these realities, but there is no reason why the title cannot be used to reinforce them, especially in maps for younger, less experienced users. Similarly, the layout of legends or the number of data classes can infer how the subject information on a statistical map has been processed and can assist in its interpretation by implying questions that should be asked by the map user.

In-between thematic and reference maps, I have identified "special purpose maps." While many of these maps seem to be thematic in terms of their titles or the small number of uses for which they are intended they do not look like the thematic map described above. The reason is that their ratio of subject information to base information is much lower. Many, as the example in figure 44 illustrates, appear as though a reference map was used as the base for their thematic or subject information. As a result, these maps usually have much more base information than is perhaps needed from a communication point of view.

Historically, there are many examples of special purpose maps (Castner,1980b). In current practice, many road maps and nautical and aeronautical charts, what Robinson (1985) calls "way-finding" maps, probably fit into this category for they often have a great deal of supplementary base information that may be of value at various times in planning or choosing a route. This might include information on places to stay overnight, access to food or services, or simply prominent or interesting landmarks along the way.

At the far right of the continuum in figure 41 is a product of the computer age. Any printout of data, often in raster form, in which a number or letter is placed within each grid cell, I call a digital map. Figure 45 is an example of such a map. It essentially has no base information because there are only the data values. The outline of the data area may simply be implied by the fact that the values end at the edge of the area. If it is a familiar shape, then the perimeter of the area with

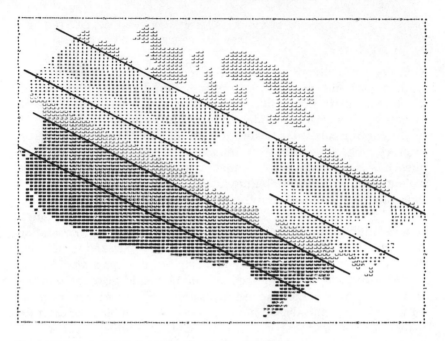

Figure 45 A typical digital map as generated by SYMAP.

symbols provides sufficient information for readers to recognize the area.

At the other end of the continuum, I have placed the aerial photograph as an image which essentially has no subject information. Since neither its graphic form nor the information it represents has been manipulated or classified, no particular piece of information stands out any more than any other, something that could not quite be said of the topographic map.

Thus, the ratio of subject to base information provides a useful diagnostic tool for evaluating specific maps in design terms. Stated a different way, and from a different perspective, identifying the activities that are proposed for a map should suggest to its designer how its specific information components are to contribute to the role of the map, namely, what ratio of subject to base information would be most effective.

One of the other products derived from the creation of the various models of cartographic communication has been the definition of the processes involved. Three terms that are particularly useful in describing nuances in map use are examined in the next section.

MAP READING, MAP ANALYSIS, AND MAP INTERPRETATION

Morrison (1976) identifies three convenient categories of map use: map reading, map analysis, and map interpretation. The first, map reading, involves the extraction of information from a map much as one uses a dictionary. It may involve identifying features and giving names to them or noting particular attributes. Map analysis, the next higher level of processing, is essentially systematic map reading involving a sequence of map reading tasks aimed at some specific goal, such as a distance measurement or finding the best route from point A to point B. In other words, map analysis involves working with information derived from map reading. As a result of such activities, we gain a certain familiarity with both the image and the place (the information represented), so that we can begin to talk about them and to describe them to others. It is then but a small step to map interpretation, the highest level of cognitive interaction. It involves adding other information that is stored in the map user's mind in either short-term or long-term memory. Typically, we might inspect an assemblage of discriminable elements and systematically match them with elements which we have learned are typical of that class of feature. Or we may consider the implication of placing an obstruction across some particular road or stream. A most simplistic interpretation might be that a particular set of contours describe a hill with a convex north slope and a concave south slope. As our experience increases, we are able to interpret some arrangements of features as being typical without any conscious matching to a "list."

Of these three categories of map use, designers of maps can really only directly influence map reading. As map designers, we have no control over the experience which map users bring to the map reading transaction nor to the time they might allocate to its study (Castner,1979a). Thus it is important to consider the process of map reading in greater detail. Fortunately, the other two categories of map use depend on map reading.

Tasks in Map Reading

Morrison (1976)[1] subdivides map reading tasks into three processes: detection, discrimination, and identification.

Detection involves the awareness that "something is there," but it has a binary characteristic in that its opposite awareness is that "nothing is there." In other words, with the onset of viewing, one is immediately aware of a general level of illumination and whether or not the paper or surface in view is blank. For the purposes at hand, it is assumed that the only case to be considered is the one where the lighting is sufficient, and the reader is examining a map.

The elements detected would be points, lines, letters, or areas of color or texture. It is likely that the only "meaning" to arise from this fleeting stage is that the image is a map and that it is, or is not, in a logical orientation for meaningful inspection. If it is not, then presumably it would be rotated until it was in the correct orientation. One can demonstrate the speed and efficiency with which this can be accomplished by simply opening a book which has no orientational clues on its cover.

Discrimination adds further information about the elements detected, although it will still be along nominal or ordinal scales. In terms of the former scale, it may be the color or general shape that is discriminated, for example, a red circle. In terms of the latter scale, it may be the relative size, value, or intensity of the element, for example, names of different sizes or patches of various colors. All of these kinds of distinctions can take place in peripheral vision. Operationally in visual search, discrimination acts to reject targets for further scrutiny by noting their failure to match one of the characteristics of the sought for object. As Beller describes the related process of filtering (1972, 137), a particular form won't be perceived as a lemon if it is red.

Identification establishes precisely complete information on the nature of a symbol, that is, its identity. To do this, the eye usually must fixate upon it. In a sense it means noting that all the dimensions of a symbol match those of the sought for element. For example, a particular object is an apple, and not a lemon, because of its particular color (red or green), shape (rounded at the top, tapered toward the bottom), size (about as large as a baseball), and skin texture (smooth).

In active map use, a map reader is constantly employing all three of these tasks. For example, in navigating with a nautical chart, a mariner is constantly comparing the environment around the boat to its depiction on the chart. Each time he glances down at the chart, he detects certain gross features which verify that the chart is in its correct orientation for reading. As the mariner seeks to find the spot where the boat is positioned, he will discriminate numerous symbols until the course line or the situation of his vessel is found. He may then use foveal vision to identify the exact position of the boat, and then to inspect some of the nearby buoys, soundings, or other features that may confirm what he has just seen across the water. The mariner repeats this sequence of processes each time he looks down at the chart. However, in this repetitive viewing environment, we can describe a fourth visual task which we can call verification.

Verification Consider the following sequence. A navigator is tracking her ship's position on her chart by visual references to a nearby buoy and a distant water tower on shore. Having detected, discriminated, and identified these three points on her chart, she will make numerous comparisons of their relative positions as the vessel makes headway through the channel. Each time her gaze returns to the chart, the general configuration of tonal contrasts and figural elements within this portion of her chart will direct her eye to the area where her vessel lies. Each time she returns to the chart, she will become more familiar with the graphic environment in which these three marks are embedded. At first she will have to re-identify through foveal inspection a number of symbols to verify their correct identities. But as familiarity with that chart environment increases, she will feel less compelled, and may need to scrutinize fewer crucial elements around her three specific viewing targets. In effect, the collective mosaic of symbols in the viewing area begins to supply the "identifying information" and thereby makes it easier and faster for the navigator to find her target symbols because she now finds them reliably in peripheral vision.

What is significant about this task is that verification starts out as a foveal process but may end up, optimally, as a process in peripheral vision. Obviously it is a process that is used more and more as one experiences a particular map image. Perhaps the speed with which one is able to make this transition from foveal to periperal vision in visual tasks is a measure of the experience of a map reader; we would associate certainly the ability to make this transition more often and more quickly with good map readers.

CONSIDERATIONS FOR PERCEPTUAL DESIGN

Now we are in a position to apply all these ideas directly to some design considerations. Briefly, the map designer, armed with knowledge of his future users and the specific activities they wish to perform with the map, sets out to create a graphic image that will facilitate the readers' use of both foveal and peripheral vision in discovering the hierarchical and planar structures in the information presented. The basic design elements – point, line, area, and letter symbols – will be given graphic characteristics that make these differentiations probable, or at least possible. The map user, in processing the graphic image, will make use of four basic visual tasks: detection, discrimination, identification, and verification.

As we have noted, the first two processing tasks can be accomplished efficiently in peripheral vision. On the other hand, the identification and verification of fine detail and the reading of labels must be accomplished by foveal inspection. For every map use activity (as defined above), each map element can be evaluated as to whether it will need to be accessed primarily by foveal or peripheral vision, or both. Taking all map use activities into account, it can be determined whether the dominant tasks involving a given map element are related to access by foveal or peripheral vision. Those that must be fixated will likely be part of the map's subject information and will need to be symbolized in some eye-catching way. Those elements that will require only occasional scrutiny will likely be a part of the map's base information and can be symbolized in some unobtrusive way. In this manner, we have a way of "allocating" information for processing to one of the two visual systems and of determining the optimum symbol form and its characteristics needed to represent those features.

One unfortunate aspect of foveal vision is the fact that it can easily be overloaded. This can produce a situation, called tunnel vision (Mackworth, 1965), in which the area of useful peripheral vision and the useful interaction of the two parts of the visual system are greatly curtailed. Such a condition would arise when the symbols chosen for a map are too large, too bold, or too complex, (there are too many attractive image elements competing for foveal attention). A good design strategy, therefore, would attempt to symbolize as many information categories as possible so that their extent or location could be monitored in peripheral vision. Innately, not all types of map symbols, however, lend themselves equally to these design demands.

Point Symbols

Point symbols by their very nature require detailed foveal inspection for their thorough identification. To make this possible, we must make sure that they are legible. This usually means making them sufficiently large in size or bold in form. For the many applications where point symbols are used to represent subject information, they must be given some prominence in the design. Of course, the more symbol attributes that are known to the map users, such as its color and shape, the easier it will be for them to detect and identify it.

However, point symbols can be designed so that they are easily detected and discriminated in peripheral vision. Williams (1967) and others (Eriksen, 1952, 1953) have shown that color is the most effective symbol attribute for making possible the search for and discrimination of point symbols. Forrest and Castner (1985) have shown that the simple step of framing pictographic symbols with some geometric shape significantly decreases their search times. For point symbols to be unobtrusive, they must be relatively small. This reduces the possibility that their internal structure or shape remains discernible, and thus limits the number of information classes that can be represented by these small symbols.

Line Symbols

Line symbols by their nature are inherently attractive to the scanning eye. The more irregular and less predictable they are, the greater this attractiveness. As a result, they are easily detected even in peripheral vision and are very effective in representing important map information. On the negative side, however, if their density or complexity is too great, it is very difficult to ignore them and search for other symbols within a heavily contoured area. (One needs only to recall a topographic map of a highly dissected landscape.) So, they are difficult to employ, unobtrusively, for a map's base information. Despite this negative aspect of their perception, line symbols continue to be used to represent many kinds of base information, such as contours for elevation and grids for location.

Area Symbols

Area symbols are almost uniquely detectable, discriminable, and identifiable in peripheral vision. The only situation in which direct foveal vision may be required to identify area symbols is when their structure is so close to that of another symbol that some spot comparisons are necessary. Even here there is potential for utilizing area symbols in more creative ways. Julesz (1975) has shown that area symbol differences can be extremely small and yet easily recognizable, although only with foveal inspection. He describes a number of textures, as in figure 46, in which an embedded pattern is discernible due to a variation in what he calls their "secondary statistics." This suggests that we have yet to begin to explore ways in which secondary characteristics or subcategories of information can be imbedded in area symbols or textures. For example, in a vegetation map we might wish to subdivide the information class of "forest," depicted in green, into "deciduous" and "coniferous" without having to utilize another hue. Thus the category "forest," with its green area symbol, remains visually a unique and unified image in peripheral vision, but with inspection two sub-

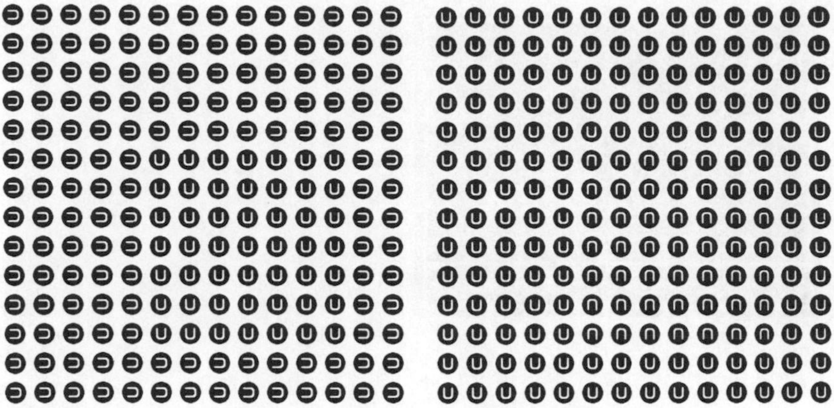

Figure 46 Two textured figures with embedded subareas: the subarea at left has a different secondary statistic; the one at right has the same secondary statistic. As a result, the embedded figure at left is not as difficult to discriminate; the identity of the one at right requires more deliberate scrutiny. From Julesz (1975,37). By permission of Bela Julesz.

categories can be revealed using, for example, two different sets of reversed letters.

Map Lettering

Map lettering, whether names or labels, almost always requires some foveal inspection for it to be identified and read. On the other hand, cartographers go to considerable lengths to provide clues as to the nature of the information being labeled so that the name need not necessarily be read. For example, we create different visual images of names on maps by varying some combination of the style, size, color, and form of the lettering used for each class of information. Just as in the case of point symbols, when several attributes of the lettering are known, it is not necessary to fixate on a name to know (from discrimination) that it is, for example, the name of a village or town. Because of the large number of typographical or letter attributes, we have a great deal of flexibility in making lettering more or less obtrusive in relation to its intended use by the map reader.

The prominence of point, line, and letter symbols can also be reduced if they are reversed out of a tonal background. In the case of lines, a much thinner solid line will remain visible in reversal in comparison to its larger black or dark counterpart. In addition, there is a reduction in the contrast between the white line and its background if that back-

Figure 47 A selection of point, line, and letter symbols reversed out of a) 90%, b) 50%, and c) 10% background screen. From McGregor (1988).

ground is screened. Several examples of this effect are given in figure 47.

Stated in this way, map design can simplistically be reduced to the determination of the information categories for each of the eight boxes in figure 48. Ideally, there might be no more than one class of information allocated to each box. Operationally, we have the technical means to add many more. But we are limited by the perceptual problems

Type of Symbolization

	POINT	LINE	AREA	LETTER
SUBJECT INFORMATION				
BASE INFORMATION				

Figure 48 A matrix of design decisions concerning the allocation of subject and base information classes to point, line, area, and letter symbolization.

created by too much information, and by the economics of adding that information. The perceptual aspect is best shown in diagrammatic form in figure 49; at some point, when the two curves cross, we begin

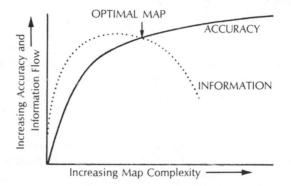

Figure 49 The relationship of information flow and map accuracy. After Jenks and Caspall (1971,243).

to overload the perceptual system and comprehension falls off.

Both of these figures bring the potential map user dramatically into the design picture. More recent interest in the market for cartographic information has further sharpened this focus on users by asking what they will be willing to pay for what kinds of cartographic information (see McGrath, 1984).

IMPLICATIONS FOR
CARTOGRAPHIC EDUCATION

With this view of the map design process in mind, one wonders how this might change our attitude toward how maps and mapping should be introduced to young children. The most important product of carto-graphic communication research has been our increased concern in map design for factors relating to the map users: their information needs; their experience; the ways in which they cognitively process graphic images; and the ways they can be misled by them. Together with this concern, we have a rich vocabulary to describe various kinds of map use and different levels of intellectual involvement. Presuma-bly, in cartographic and geographic education, we wish to train students in all these kinds of map use so that they can respond to graphic images at all levels of intellectual involvement.

However, to begin with, children do not have the experience or background knowledge that enables adults to engage in very sophisti-cated map interpretation. It would seem most ambitious that children could perform in the way suggested by this quote from a school atlas from 1922:

These maps show surface temperatures for January and July; ... isobars and prevailing winds ... the winter six months of rain ... the summer six months of rain; the annual precipitation; the natural vegetation; a relief map showing the great plains and the great highlands; and, finally, a population density map. In this series the student can *read* (my emphasis) lessons in cause and effect; for example, from hot summers and scant rain, to deserts and sparce human occupation ... In short, all sorts of geographic causal relations can be read out of this series of maps, a valuable series in the fundamentals of geography. (Goode, 1922).

Surely children can only determine causality if they have a consider-able amount of other information at their disposal. The causal relations are not self-evident in the graphic images themselves. To take a more specific example, examine figure 50. With study, it is probably obvious that there is a relationship between the occurrence of the two fish processing plants, and the fact that commercial fishing takes place in most of the surrounding lakes. However, the occurrence of the gold, silver, and copper mines has nothing to do with any other factors shown on this map or on any other map in the atlas. Neither the position of the Canadian shield nor the structural geology of this area is shown. Thus we can easily question students on where there are such mines, but not why they are there, unless we assume that the children in the

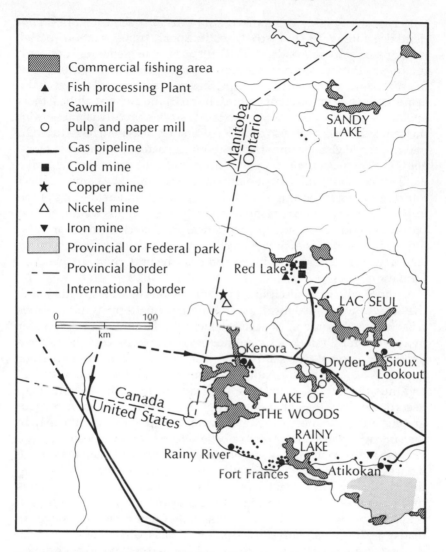

Figure 50 Portion of a map of Northern Ontario from a children's atlas. After an original in full color (Nelson,1978,28).

classroom (the map users) have the requisite experience to supply the necessary geologic information with which to address the "why" questions.

Similarly, one wonders what comparative studies will be undertaken by the average teacher confronted with another display from the same children's atlas. It compares patterns of land use for five Canadian cities: Edmonton, Vancouver, Sudbury, Yellowknife, and Moncton. The patterns are of highly generalized, and thus abstract, categories: residential, commercial, industrial, institutional, transportation, parks, main roads and railroads. A difficult lesson in generalization and regionalization would be necessary to make possible the meaningful use of the information in these maps. Since the cities are all shown at the same scale, they invite comparisons. But without some analytical tools and strategies for measuring the relative sizes, shapes, distances from the centre of the city, and so on, classroom exercises may be reduced to making crude visual comparisons about the patches of various colors. Yet there are measurements which are significant in making objective distinctions about the morphology of cities. While we cannot expect fourth graders, for example, to gain significant understanding of the impact of these patterns of land use on city life, perhaps we can at least provide ways of analysing its topological characteristics or of processing the image, that is making objective counts or measurements about its shape, extent, or arrangement, and thus the underlying mapped information.

To those who have disciplined viewing habits or who already know the answer to an analytical question, a map pattern simply verifies what is already known – it is self-evident. In the case of an unfamiliar map, we carefully follow its outline or coastline, process the relative density of symbols, or sample the place names so as to match our analysis with other information or mental templates at our disposal, and thus identify the country or geographic area. But if young map users do not have that other knowledge or a sufficient inventory of geographic images in memory, then how can they possibly use the map in the same way we do? They may only be able to say that "it looks funny."

There is no problem, however, in having children use maps as storage devices for geographic information, that is, to use them to look up things. My concern is that we probably ask them to do so much of this that it may have become an intellectual dead end. It is as though in English class we were allowed to use only dictionaries and never to read essays, poems, short stories, political tracts, directions on "how to do," descriptive narratives, and so on. What we seem not to have developed are more generic map types, appropriate not only to Euclidean but also to topologic and projective space; map types that can be used to address questions other than simply, "where is it?" The exposure of children to different classes of maps should be as rich as their exposure to the varieties of forms of the written word.

We can also do more with map analysis. Unfortunately, we have not identified and described very many procedures that we might use to examine and analyse the many kinds of images such as those reproduced in this book. Procedures which we do seem to introduce in geographic education concern the measurement of distance and area. Both of these kinds of measurements can be performed on very large map scales without generating serious errors. However, we do not warn young students of the possibilities of large errors in measurement when they perform these same procedures on medium- and small-scale maps. Certainly there is no question of the value of maps as surfaces on which measurements of distance, area, and shape can be made. But it is a conditional value, and the conditions are very difficult to explain to younger students. What we need are map analysis procedures that relate to other structural relationships and which are applicable to all map scales.

In chapter 2 I noted that many of the topological attributes of space are appreciated long before those of Euclidean space. Thus it would appear that we could develop a number of map analysis procedures for describing those topologic relationships in space. For example, some kinds of linkage analysis can show how places are connected without having to invoke the abstract conditions of Euclidean space.

In a broader sense, map analysis involves breaking down the map image in such a way that it is possible to describe it to ourselves and to others. At the most primitive level, this involves points, lines, areas, amd labels. A perceptual approach to map analysis suggests examining the characteristics of these elements first, for they are inherently simpler and easier to describe than are the classes of "symbols" delineated by the map's legend. For example, a blue line which undulates across the map tells the child more about the nature of rivers than does its "blueness" or the word "river." In photographs, man-made objects tend to be sharp or uniformly edged, often with right angles, while naturally occurring things are usually irregular in form and texture and have blurred edges Only when we have described the image can we apply, in map interpretation, other information which gives meaning to those forms and relationships noted in our analysis.

All this, I believe, evolves from a more basic problem. We tend to be teaching more about the conventions of cartographic practice than about the decisions that went into the creation of those conventions. In other words, memorizing the elements of maps is not the same as learning about the underlying information which they represent, especially that information arising out of their collective presence. As a result, we fail to introduce students to some of the more exciting problems of communication and we fail to take advantage of their innate visual perception. If we taught more of the perceptual logic of graphic images, and how they can be manipulated for specific purposes, then children could begin to use a wide range of graphics in communication rather than one small subset of them, namely, reference maps.

If we believe that visually processing graphic images in analytical, systematic, and synthetic ways is a valuable intellectual activity, and at the heart of graphicacy, then we should begin to make these skills a more central and visible feature of our initial education in maps and mapping. Perhaps then, our students in geography will be better prepared to understand and utilize the more sophisticated numerical and digital based images that they will encounter later on.

CHAPTER EIGHT

A Perceptual Approach to Geography

Even in learning a skill – painting, or music, or a sport – the learner, as he performs, must continually judge his own performance, be aware of his mistakes. Am I in tune and in rhythm? Am I watching the ball? Little children learning to walk, talk, and do a hundred other things are good at this. Too often, it is school and nonstop talking teachers that turn them into inert and passive learners.　　　　　　　　　　　　　　　　　　　　John Holt (1969, 52)

Where most researchers today are intent on transferring human knowledge directly into machine form, Lenat's programs embody such anthropomorphic concepts as curiosity, discovery, and, yes, stupidity – some of the tools through which knowledge is achieved.　　　　　　　　　　　Michael Schrage (1985)

It is my contention that one of the most beneficial products of the improvizational approach to music education taken by Orff is a sharpening of the perceptual abilities of young listeners – it is perceptual and discovery learning at its best. By increasing their discriminating abilities for auditory material, children are able to detect and appreciate more subtle attributes and contrasts in that material. In addition, by the very nature of their participation, children are provided with the varieties of feedback that are so useful in judging one's performance and thereby promoting a greater *understanding* and thus enjoyment of the music variables involved. As a result, they are also able, with increasing confidence, to deal with other more complex material, whether it is in music or spoken language. I suspect that in the long run, they also find more satisfying the more subtle and complex rhythms, harmonies, dissonances, and structures within classical music.

The question that I raised in 1980 (Castner, 1980a, 1981) was whether or not there should or could be a similar educational method for visual perception and discrimination. This would appear to be an important goal for art education. But my impression then, as the analysis of chapter 5 suggests, is that it is not a goal of art education or if such a method exists, it has yet to be promoted widely. As worthy as developing a sense of aesthetics and an ability of self-expression may be, the setting of art education, for me, is not as rich as it could be if there were a more visible component concerned with graphic communication and visual discrimination. If no leadership in this direction emerges from the ranks of art educators, then perhaps it is up to geographers and cartographers to take action. In some respects, we are in a better position and have a greater mandate to explore both the analytical and the synthetic uses of graphic images. On the other hand, there is a danger that graphicacy, as an educational concept, has been too narrowly defined in a traditional geographic-cartographic framework. In other words, the plea of Balchin and Coleman (1965) that graphicacy should join articulacy, literacy, and numeracy in our basic education was not inclusive enough.

This chapter identifies a set of inclusive, and thus unifying, concepts that describe what geographers are trying to discover in the world – the component concepts on which they build their generalizations about reality. These concepts have been developed through an examination of the attributes of reflected light discriminated by the visual system. In other words, I have attempted to build an educational approach to geography based on the skills of visual discrimination – skills which are just as fundamental as those used in auditory discrimination and those used in reading, writing, and arithmetic. It should be demonstrated to politicians, educators, and parents that "going back to basics" (or forward to fundamentals, as one astute bureaucrat put it) should include these two essential perceptual skills – listening and looking. If they are not already evident, perhaps it would be useful to outline, at this point, some of the advantages of a more improvizational approach to discovering the world through vision.

ADVANTAGES OF A
PERCEPTUAL APPROACH

Information from and about the world is the basic ingredient from which maps are made; the processing of that information can be done either directly or indirectly. We do the former when we witness the world around us through our senses, primarily our visual system of eyes and brain. Indirectly, we can process some representation of it, whether that representation is numeric, analogic, or graphic. But even when we utilize some representation of the world, we process it through vision. Thus the eye-brain is the most relevant system we have for discovering information about the world.

Why then do we not take advantage of this knowledge in introducing the world to young children when they reach school? It is my contention, and the evidence abounds, that we have allowed the teaching of geography to be reduced to the study of our various representations of the physical characteristics of the world rather than the information they represent, namely the characteristics of maps rather than geographic information. In other words, we teach about map products rather than about the processes of mapping which broadly speaking include: thinking about some aspect of the world, seeking its relevant dimensions; establishing a communication goal, considering the various modes and forms of its possible representation, and only then, executing that representation.

By beginning with visual perception, we can start working with the child's own visual experience of the world, as inarticulate as it may be. By school age, most children can move about successfully in space. This suggests that they have worked out most of the basic perceptual problems that relate to objects and surfaces in space, and to some of the primitive topological relationships between them, even if they cannot verbalize them. Thus we have, in essence, a concrete and shared point from which all lessons can depart.

In addition, the earth provides an infinite number of settings for study, hence it is our richest source of curriculum material. To use it initially, we must access it through the most available and most personal (and thus the most meaningful) way – the sense systems of the children. For geography this means primarily (but perhaps not exclusively) the visual system.

Utilizing the environment as an object of study provides a third advantage, one that is primarily tactical. Besides its infinite variety, the environment is complex. As a result, most questions we can ask about the world can only be answered in a relative way, that is, they can only be better or worse for some purpose, not "right" or "wrong." Measurements, in this context, have potentially greater educational value in their support of arguments than in anything intrinsic in themselves.

Thus an educational approach that seeks questions with more than one possible answer is one that leads naturally to an examination of the criteria by which one answer is judged to be more satisfactory than another. In cartographic communication, this means determining the specific activities for which the mapped information will be used; in cartographic design, this means considering how visual tasks will be used to process the graphic image. Teachers in this general context become helpers in making measurements, in arbitrating the admissibility of evidence (measurements), and assessing the logic of arguments.

More specifically, this means teachers must be able to articulate (1) the specific purpose for communicating an idea or piece of information, (2) the appropriate level of scaling of the measurements to be taken, and (3) the graphic and intellectual logic of the arguments to be pursued. This whole approach is thus related to the utilization and development of critical thinking skills on both sides of the brain.

All this produces an additional educational advantage. Even the best answers to a question will inevitably be flawed in some way. The imperfection of any answer thus provides the opportunity of re-addressing the question at a higher or more complex level of measurement or analysis, or at another level of generalization. Readiness in this educational situation would thus be defined as the child coming to an understanding of the imperfection of a problem's solution and an appreciation of the need, or possibility, of another, more sophisticated, approach.

There is an analogy to this process in vision (see figure 35, p.125). Stored information on the nature of the world directs our perceptual exploration of it. Through our sampling of actual environments, we verify our understanding of them but also turn up new or contradictory information which in turn modifies our stored image. And so it goes in repeated cycles over the short-term experience of that environment. It makes sense that our educational environment should provide for such cyclical experiences because every time we examine a concept or setting, we see it in new ways because of what we have learned about it before.

In other words, few concepts of any significance are mastered in one lesson at some particular grade level. Rather, we must keep coming back to them so as to see them in greater detail and to apply them in new ways and forms to more complex problems and circumstances. To foster such a spiralled approach to the curriculum in geography we need to isolate those concepts that can be elaborated repeatedly as the child's ability to handle intellectual complexity and abstraction increases. It seems to me that the most useful concepts will be those related to the processes of mapping, not the products of mapping. Thus, for example, having children in the elementary grades memorize the names of those intersecting lines on the globe and the cardinal directions is not teaching them anything about the process or concept of geographic orientation.

Finally, the abilities to process words and numbers are touted as the most fundamental and powerful skills for survival in our society. And indeed they are important. But they represent only one way to process information – a way that is analytic, symbolic, abstract, sequential, rational, logical, and linear. We have come to think of these as processing characteristics of the left brain or, perhaps more correctly, of left-mode thinking. But the nonverbal, synthetic, concrete, analogic, nonrational, spatial, intuitive, and holistic characteristics of right brain or right-mode thinking are just as valuable.

To pursue this briefly: it is regrettable that there seems to be a trend toward stripping the educational curriculum of places for right brain learning to take place as we reduce not only the availability but also the richness of programs in art, music, and physical education. Thus geography may well be one of the few curriculum areas left that can provide activities that develop both sides of the brain; that can provide experiences with both classes of information processing skills – it is not just a left or right brain discipline! Our viewing a graphic image or an environmental scene is an extremely complex action involving a variety of visual and intellectual skills. It is one that is achieved with little apparent effort, but one that has many subtleties.

WHAT HAPPENS IN VIEWING A MAP OR SCENE?

In recent years, research in cartography has focused on the ways in which people interact, both visually and intellectually, with the basic scenes and images that map makers and geographers utilize in their work, namely, landscapes, photographs, paintings and drawings of landscapes, and maps of various kinds. These interactions raise questions about both the cartographic information involved and the ways it is to be used. Further, we realize that we come to understand the world in three interactive and educational ways (figure 32, p. 119): by looking at it, by thinking about it, and by trying to represent it. Or, as McKim put it, by seeing, imagining, and drawing. It is significant that all three of McKim's words are verbs which we can easily associate with processes or educational actions that we might take in making discoveries about the world, in what we might term "geographic thinking."

It is tempting to replace the word "drawing" in McKim's diagram with something else, like "mapping" or perhaps "modelling" (Carswell, 1986). But maps are not the only images we create to present our ideas. And while we may have been too quick to drop landscape drawing from our geographic curriculum, we need a word which incorporates all of these ideas of "attempting to represent graphically the essence of the reality at hand." Another approach might be to modify McKim's diagram in response to the analogy to vision mentioned above. If we did, a model of geographic thinking could reflect this cyclical nature of perception and our dual sources of information about the world,

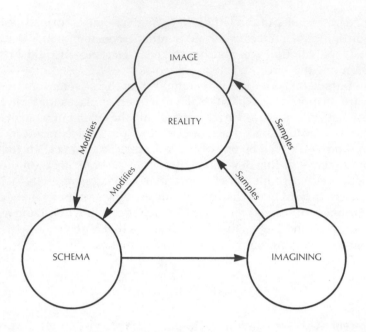

Figure 51 A model of geographic thinking showing the cognitive interactions between memory, imagining, and looking at reality, whether the latter is seen directly or in images.

namely that which we witness directly in the world and those that we process indirectly from images of it. Figure 51 suggests how these ideas might be represented. Strictly speaking, the image referred to in this figure might be a map, but equally it might be an aerial photograph, a landscape painting, a sketch map, an architectural plan, or any of the images implied in figure 41, p. 143. As well, the schema represents whatever information we hold about a given geographic place or concept. This knowledge is manifest in (1) the images (or mental maps) we conjure up in our perceptual exploration or imagining (to use McKim's term) of the place or concept, and (2) our visual sampling of images and scenes. The only partial overlap of the image and reality circles reminds us that our images are never perfect replicas of the realities they represent.

It is always fascinating to watch and listen to an expert describing what appears to the novice to be a simple and mundane scene or image. The expert is able "to see" so much that escapes our unpractised eyes. Lynch (1960,123f) documents the richness of environmental images held by primitive and advanced people alike. Major Peniakoff, operating in the Libyan Desert in World War II, describes the great skill of even

Figure 52 An aerial view of a landscape .

young Arab children in finding their way across the flat, stony expanses where there were apparently no conspicuous landmarks.

...landmarks, so minute as to escape our notice altogether, remain fixed in their memory; not only visual landmarks but also the feel of the ground underfoot. Moreover they have in their mind a picture of the general structure of the country: on a featureless plain, they can distinguish watersheds and slopes. Dry, sandy watercourses faintly marked with poor bushes seem to us distributed haphazardly; the Arab recognizes them as part of the general drainage system, flowing ultimately to one of the big wadis. (Peniakoff,1950,162)

Oh to be able to give our students such training in perception! But of course an experienced geographer uses similar skills in analysing maps and photographs. Careful analysis of an image like figure 52 allows for the identification of various image elements, such as the kinds of trees or types of buildings, but also it can lead to ideas about more subtle things, like the standard of living, whether the area might have been glaciated, what the culture of the populace is, and perhaps even some speculation about the exact geographic location of the place depicted.

   The identification involved in such analyses depends upon shape recognition and the discrimination of certain attributes of small features and areas. It is a skill largely reducible to a set of directions. The interpretations, however, are not, for they involve matching certain arrangements of elements or textures in the images with those held in the geographer's memory. The information required for this may not actually be in the photograph. While the expert interpreter may never have seen the photograph before or been in the place depicted, it undoubtedly held invariant combinations of elements and features that matched those in memory. In a sense then, the photograph acted as a lightning rod for the knowledgeable geographer, allowing him or her to recall those associations from memory and to talk about the place depicted.

But you and I may not be able to perform the same interpretative feat if we do not have command of the same set of information in memory. On the other hand, part of what the expert did, and it probably went undetected, was to look for certain kinds of things and perhaps to look for them in certain ways. Thus, there were aspects of the photographic image which were more meaningful to him than to us. Once they are pointed out to us, then we too have a greater chance of looking with meaning at scenes and images, including photographs and maps.

While many research cartographers have become interested in the ways in which people interact visually with images, scenes such as figure 52 may also be of particular interest to artists and historians, to name but two other disciplinary groups. The artists may wish to describe the scene as it is or to inject a particular emotion or philosophy about the world into their interpretation. The historians may wish to seek out the artifacts of historical events that may be present. My geographic conditioning leads me to assume that historians might actively search this landscape as though it were a palimpsest of previous human activities, values, and attitudes toward the land.

In any case, geographers are not the only ones with an academic interest in the information about the world that may be available in scenes such as this. As a result, any suggestions that I might make for changing our approach to geographic education may be claimed by others. But my purpose here is in discovering the commonalities of interest among these disciplines rather than in establishing academic jurisdictions, particularly at the lower school levels.

PERCEPTUAL LEARNING

Since infants do not arrive fully equipped with knowledge and strategies to deal with the complexities of the world, there have to be ways by which they begin to gather knowledge about the environment. Perception is that process by which this information is gathered first hand. Perceptual learning, then, refers to an increase in one's ability to extract information from the environment, as a result of experience and practice, with stimulation coming from it (E.Gibson, 1969, 3).

In the case of viewing scenes and images, practice and experience allow us to "see" more and more in the image, that is, each time to discriminate finer and more subtle attributes of the image. From viewing them, we can make more subtle interpretations of what we see. These interpretations can be at various levels of sophistication, from the everyday adjustments of putting on spectacles to the highly professional ability of chick sexing (E. Gibson, 1969, 5-7). An optic array, then, carries information which is found by analysing and discovering its structure over time. The information, however, is *potential*, for it may not necessarily be picked up by the perceiver, or perhaps is only picked up at certain levels. Thus perceptual or discovery learning could take place with quite complex visual structures as long as our expectations about what we are seeking are in themselves graded.

Resnik (1976, 76) considers that the real problem in education is not the choice between teaching by rules or teaching by discovery, but of finding teaching rules that will enhance the probability of discovery. Donaldson goes on to develop (1978, 106-20) a number of arguments for making use of errors that normally and spontaneously arise from inquisitive and active exploration, and for the unnecessary need for external rewards in learning. In the end, what children work out for themselves by discovering the inadequacies of various theories and of inventing better theories will have a quite different status in their minds than what is told to them by an authoritative adult. Thus children may be able to experience the truth of the matter and be their own judges. When they are unable to see for themselves, that is, when their experience is insufficient, then they must be aided and helped to face and overcome their errors without feeling defeated or wanting to withdraw from the learning activity.

To facilitate the learning of a graded set of information or structures in a visual display, it would seem necessary that we provide some kind of organization whereby structural elements or components can be discriminated and identified, in other words isolated and named. One source of organization is in vision itself. As I have proposed in both chapter 3 and 7, there is a bipartite aspect to vision that suggests that we process images in two essential ways. We discriminate gross features, areas of distinct color or texture, edges and gradients, and a variety of other structures within an array of reflected light. From such isolated elements, we can then focus our attention on specific ones and describe or identify them in particular ways. Thus we can speak of image characteristics which we discriminate and image characteristics which we identify.

In the simplest case, an analogy was made in chapter 4 with music. The four basic attributes of sound – pitch, loudness, timbre, and interval – were parallelled with the corresponding attributes of light: hue (wave length), value (brightness), chroma (purity), and texture (pattern). An understanding of these dimensions, and an ability to discriminate variations among them is fundamental to developing one's visual perception (Castner, 1981, 65). Thus our lessons in cartography and geography might rightfully begin with considerations of the fundamental perceptual dimensions of color and texture.

A PERCEPTUAL APPROACH
TO GEOGRAPHY

A perceptual approach to geography rests on the proposition that the basic discriminations we make in visual perception can be related to a similar number of cognitive identifications of categories of geographic information. An analysis of these categories, in turn, suggests a set of defining processes or conceptual goals. These are the essentials of geographic thinking, no matter which systematic aspect of the discipline one is considering. In other words, the processes are common to both physical and human aspects of geographic inquiry.

By recognizing that these categories of geographic information can be actively sought for in the ambient array of light, we provide the basis for a more dynamic, discovery approach to geographic education. It is equally clear that some of the sought for information is explicit and some is implicit. By explicit I mean some is related to the existence of objects, places, boundaries, and so on, while other information which springs from the topological and Euclidean relationships among these explicit features is implied or implicit. Geographic education should, therefore, provide for experiences in looking for both explicit and implicit categories of geographic information. Perhaps by reducing the emphasis on the wealth of explicit geographic information, we can direct more attention to searching for ways to discover the implicit information in both landscapes and images. If this interaction between processes in vision and geographic thinking is as I have suggested, then it points to the fundamental importance to geographic thinking not only of working with images and representations but also with and in the environment itself (field work).

Each of these assertions is presented in diagrammatic form in figures 53, 54 and 55. Each of the related discriminations, identifications, and defining processes is placed in consistent positions in the array of circles and, momentarily, will be described. But first I will briefly describe each diagram.

DISCRIMINATIONS OF:

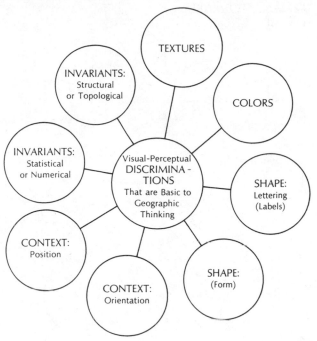

Figure 53 Eight perceptual discriminations that are basic to geographic thinking.

The first diagram, figure 53, relates to the nature of the visual system as described in chapter 3 and manifest in the terminology developed in chapter 7. In Gibson's terms, the circles relate to the characteristics and structures embedded within ambient light, namely, in light reflected to us from the surrounding environment. They relate to the general categories of visual stimulus dimensions that are discriminated, but not necessarily identified, in our visual field of view, especially in peripheral vision and without the need for our conscious attention. The products of this particular kind of thinking are different from those we identify in figure 54 in that they are less specific. This may be because we have not paid much attention to them, or they are in fact harder to isolate. The very nature of visual discrimination involves holistic processing in which features and elements are less important than textures, gradients, and various Gestalt properties, which may be shown, are our best evidence of particular kinds of relationships – the implicit information in the environment. As such they apply equally to light reaching us

FROM THE DISCRIMINATIONS WE
ARE ABLE TO MAKE
IDENTIFICATIONS OF :

Figure 54 Categories of geographic information that are basic to geographic thinking.

from both dynamic landscapes and images, whether it be from our movements relative to them or their movements relative to us as observers. This interrelationship alone speaks to the essential unity of geography and cartography – of geographic thinking and mapping – for we react *perceptually* (but not cognitively) in the same way to landscapes and the representations we make of them.

From these discriminations, we are attracted, in various ways, to individual elements in our field of view. It may be due to their unusual nature or strong contrast with their background. It may be that we are looking for some meaningful characteristic of a sought for target, or it may simply be due to curiosity. In any case, we are able to identify specific things, whether they be features, labels, regions, or some kind of relationship, context, or structure. Having identified such things, geographers and others go on to study and categorize them in various ways.

THE ESSENCE OF GEOGRAPHY
LIES IN DEFINING:

Figure 55 The conceptual goals of geographic thinking.

In the second diagram, figure 54, the circles represent basic categories of geographic information. They are the kinds of things that we teach about geography – the explicit information about places in the world, and features recorded on maps. They are the products of a particular kind of directed and usually conscious thinking. As such, these categories (with the exception of the two on invariants) form the content of our traditional approach to geographic education – one which may over-emphasize memorization of such explicit information as "capes and bays," state capitals, conventional symbols, cardinal directions, and so on.

In some ways, the categories in figure 54 are similar to the five themes in geography recently described by the AAG and NCGE: location, place, relationships within places, movement, and regions (AAG and NCGE, 1984). However, my categories are more fundamental or primitive. As such, they can be defined more precisely, can be broken down into more manageable subconcepts, and can be treated at a variety of levels of measurement and generalization. As a result, they may prove to be more valuable in classrooms and as guides in the development of analytical procedures and generic types of maps noted in chapter 7.

In order to prevent both these categories and the above-mentioned themes from being treated simply as inventory items in curriculum, they must be placed in a more dynamic educational context. This is accomplished by relating them to figures 53 and 55.

Figure 55 represents the logical extension of figure 54 if we apply to its components the defining processes of geographic thinking. In essence the circles represent the conceptual goals or products of geographic inquiry. By describing them in this way we can increase the emphasis on a more active discovery approach to our world – one which is goal-directed in terms of finding a defensible solution to a question of analysis or communication. In a sense these goals are in themselves concepts or themes around which we can also define our discipline.

In order to link these three diagrams together conceptually, let us consider the derivation of the eight circles and then describe each of them individually.

There are five attributes of reflected light that our eyes are able to discriminate without close scrutiny: color, texture, shape or form, a context, and invariant structures (or simply invariants). These, collectively, provide an initial framework upon which we begin our visual inspection of a new scene or image. These global elements form the visual structure in which subsequent fixations will make the identifications necessary to solve a problem or reach some goal. It should be stressed that the emergence of this initial figure-ground arrangement takes place spontaneously, although it remains accessible throughout the viewing.

However, we are not equally sensitive to these five attributes. For example, most viewers do not normally discriminate independently the three dimensions of color, so that the attributes of hue, value, and chroma can easily be considered under the single heading of color. On the other hand, the discrimination of a "context" and "invariants" is so richly complex that it can be elaborated in many ways. There are also those special shapes and forms we associate with alphabets and language – the names and labels we give to places. In the end, therefore, I have settled on the eight perceptual attributes, shown in figure 53, as the basis of an approach to the conceptual goals of geography and for a discovery approach to elements of geographic thinking.

I make no claim to the mutual exclusiveness of these attributes. When considering vision, or indeed any of the other sensory systems, there are so many interconnected and overlapping associations and linkages that few objects or surfaces can be considered to have perceptual dimensions that are truly independent. Briefly, the attributes can be described as follows.

Textures

In a sense, all images can be broken down in perception into areas of texture and color as well as into shapes, lines, or edges of various kinds. There are two levels at which we respond: (1) to individual areas of homogeneity made up of relatively similar elements and (2) to assemblages of quite different elements which are seen collectively.

Examples of the first level can be seen in any terrestrial scene where distinctive textures define individual features, the hardness or nature of their surfaces, or their position and spatial attitude relative to ourselves. Textures that change in the form, arrangement, or orientation of their constituent elements are generally perceived as different in kind or quality. As a result, they convey qualitative information about the areas or surfaces from which they are seen to arise. In map making, we use such textures to identify taxonomic regions (for example, kinds of soils). Other textures, which change in the density, size, or spacing of their constituent elements, are also seen to be different, but the impression of difference is normally in terms of importance or prominence. As a result, they convey quantitative information about the areas from which they are seen to arise. We use them in maps to identify hierarchical regions (for example, amounts of rainfall). These are very primitive perceptual skills which are probably mastered if not articulated at a very early age.

Examples of responses at the second level include the spontaneous recognitions that one is looking at a particular type of image (a photograph as opposed to a topographic map), landscape (an urban or rural scene) or perhaps a particular map scale (a large-scale urban plan as opposed to a small-scale regional map). In such cases, viewers are responding to the scenes and images as arrays or mosaics of individual elements which form rough textures and which they may have learned to associate together.

In order to make assignments or associations of meaning to homogeneous areas of texture in scenes we view or maps we make, viewers must be able to detect some systematic uniformity in the six perceptual dimensions of textures: the type of textural element(s), their size, their spacing, their arrangement, the orientation of that arrangement, and their collective impression of pattern-value (Castner and Robinson, 1969). Where we detect changes in some of these dimensions, then we have evidence of the existence of a zone of transition between two areas.

Depending on the level of generalization we are considering, these transitions may be considered as meaningful boundaries.

However, the essence of geographic inquiry lies in the isolation and discovery of boundaries between hierarchical or taxonomic regions. We often represent these on maps as distinctly different regions by giving them separate colors and distinct, solid boundaries. But in reality, such distinctions rarely occur. Rather we find vague transition zones from one area to another in which the regions are defined not by some single factor but by a change in the weighting of a multitude of factors. In a sense then, the regions are detected by the discrimination and identification of subtle changes in textures or mixtures of component elements.

Thus the search for such subtle changes in the perceptual dimensions of textures and ways to describe them may provide more intellectually exciting activities in school than by giving students an overdose of previously defined boundaries. Obviously we make simplistic statements about areas and regions in order to speak more efficiently about them. But unless students (and their teachers!) understand the level and nature of those generalizations, and how they were derived, the statements will be meaningless and sterile. Basic to this ability is an understanding of and the capacity to describe the six perceptual dimensions of texture.

Colors

In a similar way, we discriminate differences in the three perceptual dimensions of color. From them we are able to identify surfaces, objects, or areas. In a sense then they are regions of homogeneity to which we assign names or meanings. As with textures, changes in certain dimensions from one area to another are associated with particular kinds of changes. Changes in hue are usually interpreted as differences in kind or type; changes in value as differences in magnitude. We can make use of these principles in constructing area symbols that identify taxonomic or hierarchical regions on maps.

The useful application of color perception in geography requires improving one's ability to discriminate the three basic components of color: hue, value, and chroma. Given my discussion of music in chapter 4, this suggests developing an improvizational way to work with and manipulate colors. The use of colors in design work necessitates the ability to reduce the near infinite number of possible color choices to a workable few, and to appreciate how the effect of simultaneous contrast may work to confound our efforts in selecting an easily discriminated set of colors. For both of these abilities, it is useful to have a perceptual model of color space and strategies for sampling and manipulating that space.

Shapes, Forms, and Lettering

The discrimination of shape or form allows us to detect point or linear objects on the earth or similar features on a map. To do this we respond to small isolated areas of color or texture. To identify points or linear objects, we must examine such primitive dimensions as size, number and relative length of sides, number and regularity of corner angles, color, internal structure, and so on. In maps, we call these elements point and line symbols.

As with the determination of regions and their boundaries, the choice of a logical and useful symbol involves an understanding of what it is we are trying to represent. Like the representation of regions, there are many point and line symbols which have been created and are available for use and study (but not necessarily memorization). But learning some symbols does not necessarily teach the logic behind their creation. Perhaps it is obvious enough that pictographic symbols resemble, at some level of abstraction, the objects they represent. But other symbols, common in shape but different in size, are used so that they can convey information about changes in magnitude. Still other symbols, common in size but arbitrarily different in shape, are used to convey qualitative distinctions or differences in kind. Thus there is a logic to symbol selection and design which relates not only to the nature of the object represented but also to the symbol's role in specifying the classes of information being mapped.

When we come to deal with much of the content of modern geography, we often find it expressed in numbers. But like objects that we caricature in pictographic and concept-related symbols, data sets have internal structures and descriptive dimensions that can be used to characterize them. Thus the process we call symbolization is essentially the same whether we are looking at objects or at information. In geography, the critical skill is learning how to isolate and describe the essential character of those phenomena we study, whether they are objects, groups of measurements, or counts (total number of things counted). Because of the abstract nature of numerical expression, the search for the expressive characteristics of data sets is considered in a separate circle.

Designing symbols is perhaps the most basic operation in both art and cartography. In essence, it can be the arbitrary assignment of meaning to some kind of mark, whether it be pointal, linear or areal. But the logic of the process is found in matching the intellectual or structural qualities of the object or data with appropriate visual dimensions of a graphic mark. This means utilizing the various perceptual dimensions in some systematic way. Thus our primary educational focus should be on drawing and designing symbols in which use is made of the various perceptual dimensions of shapes and lines to provide the viewer with some kinds of meaningful associations between the symbol and the information referent – the basis of identification.

There is a special case of form perception which involves the discrimination of the letters of an alphabet and the identification of words. A casual perusal of a map quickly allows us to know where there are names and labels even if we do not attempt to read them. There is something about their shape, size, and texture that makes them distinctive in peripheral vision. In reading text, it has been shown how few of the individual words actually need to be fixated upon and, similarly, how each letter in a word need not be acknowledged. While individual place names on maps lack the grammatical context and redundancies of language, they have their own contexts which undoubtedly allow viewers, reasonably familiar with the geographic context, to identify names without actually scrutinizing them.

There are two ways in which we use words in geography: to provide features with their geographic meaning (city, esker, plateau) and to indicate their cultural meaning (their given names such as New Amsterdam, St Louis, Suomi). The placement of these two kinds of labels on maps requires an understanding of some of the laws of Gestalt psychology and thereby presents another improvizational opportunity for manipulating visual variables.

The former use of words involves the description and interpretation of both the physical and cultural landscapes. These words make up our geographic terminology. The latter use is concerned with the way features are referred to, either generally (geographic names), or by the people living there (local names). It can be argued that local names, that is, the names used by people living in an area to describe their own region, provide the richest possibilities for geographic education; they furnish an introduction to the language, and thus the culture, of the area under study.

There is another aspect of language that may also be appropriate to geographic education – the sounds of particular countries or geographic regions. They may be of two types: sounds of the indigenous and descriptive music, as noted in chapter 4, and those of the language used there. While geography teachers should not become music and language teachers, we should be sympathetic to these aspects of culture and thus geographic environments and take steps to expose our students to them and to encourage their interest in and respect for the music and language of others.

Invariants

It was suggested above that there is a parallel to the process of symbolization in the description of sets of numbers by means of other numbers. Formally, descriptive statistics is involved with these matters by providing techniques that help us to make generalizations about data sets and to discover structures within them. These structures, in particular, convey information about internal relationships among the constituent members of that data set and thereby determine their unique character.

It is obvious that the quantitative revolution and the computer have made it necessary to consider more carefully these numerical procedures for they are basic to the discovery of many of the kinds of boundaries and regions that geographers are looking for. As evidence of this, most university geography departments have their own required courses on statistical applications in geographic research. However, it is a moot point whether students in those courses really achieve an understanding of the underlying principles involved. I say this because too much is asked of in these courses in terms of content, and most students are looking for specific procedures and applications and not for the fundamental ideas that would actually make the procedures easier to apply. To provide a place for these underlying concepts, I have identified one of the circles as statistical or numerical invariants.

In the same way, I believe that there is a significant and useful parallel in vision, where, relationships and structures can be detected in images and provide basic geographic information. From the materials reviewed in chapters 2, 3, and 7, it would appear that many of these relationships are topological in nature and that they are relatively innate in perception. However, in vision, we have not developed nearly the same number of analytical techniques that we see in statistical analysis. Thus I am suggesting that the discrimination of invariants leads to the identification of all sorts of associations, connections, and relationships. What invariants are and how they are identified is perhaps at the heart of all geographic inquiry. While the question of whether invariants can be systematically discriminated and identified is yet to be demonstrated in any formal way, there is no doubt that we do. I have called the key procedure that would be invoked in geographic education for detecting the invariant dimensions and structures in graphic displays "visual map analysis."

Context

In looking at a scene, such as the one in figure 52, we are immediately aware of its scale and of our orientation to it and to the objects or surfaces within it. In other words, we see the scene in relation to ourselves; its perceived perspective geometry includes each of us as viewer. This would also essentially be true of any map we might view which has a recognizable level of generalization (a topographic map or atlas map) or defined area (the outline of a continent, country, or province).

This latter situation begs the other contextual question, one that is independent of how I see myself in terms of a pictorial or other graphic image. This question asks: "Where is the place that is represented in the image?" This is a question of position on the surface of the earth; the former involved the question of its orientation, both to me the observer and to the objects in the place or scene.

Context, then, has two somewhat independent aspects that relate to (1) position (which most might agree is primarily in the domain of geography) and (2) orientation (which certainly is not, for both art and music education, to name but two, have important contributions to make to this aspect). Under the former, I would include the study of map projections and coordinate systems, such as the UTM grid and that of latitude and longitude. As such, the context of position is a somewhat abstract topic especially for younger students. This does not have to be so, but teachers and curriculum designers have been reluctant to inject bridging lessons on such things as the simple transformations that can be observed everyday around us. In contrast, the context of orientation is a most concrete phenomenon because it is something we experience almost as soon as we open our eyes in infancy. Therefore, it involves the assessment of the position of objects relative to ourselves and of ourselves relative to others. It also involves developing ways by which we can express those relationships, namely through the use of various kinds of landmarks. Certainly there is an abstract level at which we can precisely describe directions and orient ourselves, but it is not necessary that we begin our studies of orientation with those abstract systems.

Thus the goal of geographic thinking is to define a context of position and one of orientation that are appropriate to the scale of the study being conducted and to the level of generalization required to communicate information to students at a particular level.

CONCLUSIONS

There are some inconsistencies in the content, overlap and placement of these particular circles. I have already mentioned the similarity in the

CONCEPTS OR ACTIVITIES WHICH
MAY REVEAL THESE IDENTIFICATIONS:

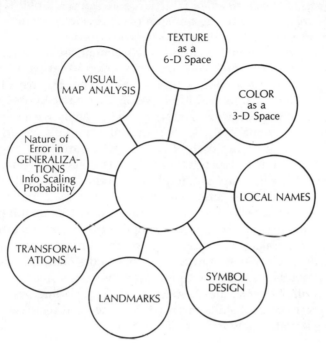

Figure 56   Eight concepts or activities which may reveal the conceptual goals of geographic thinking.

processes of symbol design and descriptive statistics. Place names and geographic labels apply only to images, such as maps, where we have labelled things, rather than to landscape scenes and photographs. The two classes of invariants, besides being objects of study in their own right, are also used in determining regions which are then symbolized in maps by textures and colors. Thus, the search for invariant properties is one way we determine the existence and nature of regions. By discovering some kind of typical homogeneity or some kind of significant change, geographers are able to draw in boundaries around our geographic regions.

There may also be some other realizations that will come out of structuring our discipline in these diagramatic ways. For example, it may be useful to think about such notions as "place" as a special invariant dimension. We have all experienced the recognition of a place that we have seen or been to before, whether it was yesterday or decades ago. From this we can postulate a non-metric sense of place that arises out of our recognition of some set of properties about that place which may be invariant and which we remember.

In terms of "teaching materials," large scale topographic type maps are probably some of the best images that provide educational opportunities in most if not all of these circles. Aerial photographs and landscape drawings of the outdoor environment, are the other obvious classes of materials. But there are other images that we have found useful in cartographic communication that do not involve all of these circles and the concepts they imply. While each circle may suggest the nature of some specialized diagram or image, I only wish to point out that an improvisional approach to geographic thinking would involve, at first, working within each of these concept areas or with images whose logic or use depends only on a few specific concepts. For example, a "linkage map" is a simple, nonmetric drawing which shows how a group of cities are connected together by a transportation system. The linkages can be simple straight lines that do not preserve, at scale, the accurate lengths and shapes of those connections. Thus lessons using linkage maps can focus on the connectivity relationship among those cities without the need (or the distraction) of having to worry about their exact geographic locations or the directions between them, until, of course, these aspects become important. As analytical and representational skills improve, and the need increases to more closely simulate real world problems, more concepts or conditions can be included.

Despite the inconsistencies among the various circles, I believe it is possible to derive from the conceptual goals of figure 55 an educational agenda in geography that can provide both the excitement of intellectual discovery and experiences with the underlying concepts of geographic thinking. Toward this end, I have developed elsewhere a considerable amount of materials demonstrating the richness of each of these topic areas as they might be elaborated in classrooms. Each discussion begins in a most concrete way, so as not to exceed the likely experiences of young children, and is followed by various elaborations to the more complex and sophisticated levels that are of value to adults. One such educational agenda is embodied in figure 56. Each concept or activity circle suggests a focal point or theme that can be explored in improvizational ways so as to discover the fundamental concepts of geographic thinking. Some have already been developed and tested in classrooms; others await elaboration. It is hoped that this volume will help to provoke both thinking about the concepts that are fundamental to geography and to suggest ways in which they might be manifest in geography classrooms.

Notes

INTRODUCTION

1   For example, Petchenik (1983) provides a thorough analysis of the mismatch of assumptions and expectations in academic research and the world of practical design. Some different perspectives are offered in Castner and Cromie (1982).

CHAPTER 1

1   Proctor (1984) argues for a fifth communication skill, physiognomacy. Its productive element involves movement, dance and expression; its receptive element involves watching and interpreting. In chapter 4 I will consider the role of movement in the context of music and the development of spatial awareness.
2   Blaut (1987) suggests that young children can gain the most from and enjoy best complicated maps even if they do not fully understand them.
3   Thorndike (1950) also noted a similar division and pointed out the need for children to be able to apply their understanding of general problems to the solution of specific problems.

CHAPTER 2

1   Concise summaries of Piaget's work can be found, among other sources, in Hooper (1968), Beard (1969), Samples (1976), Singer and Revenson (1978), and Brooks et al.(1983).
2   Most recently, one can read calls for mapping at earlier ages in such works as Boardman (1985b), Catling (1978), Blades and Spencer (1986), and Matthews (1986).
3   In fact, there are visual tasks more basic than these that must first be

mastered. They include aiming both eyes at a target (alignment); smooth pursuit movements in tracking a moving external object; binocular depth perception (localization); internally controlled saccadic eye movements, as in following textual material (reading); and spatial orientation and transformations (visualization) (Parrott, 1986).

4 In Ontario, there has been a call for the return to province-wide Grade 13 (final highschool year) examinations. The reason cited is that between 1965 and 1981, the number of Ontario Scholars, students who had achieved a grade average of 80 percent or better, had increased from 7 percent to 27 percent of all graduates. In offering an oblique explanation for this phenomenon, one member of the educational establishment noted that during the same period, many Olympic records had been broken, yet no one had questioned the stop watches!

5 Perhaps only in the writings of Barbara Petchenik, in describing her work on the *Atlas of Early American History*, can we find some of these problems discussed in appropriate breadth. See, for example, Petchenik (1977a) and (1977b).

CHAPTER 3

1 To detect the blind spot, place a small black dot or *x* in the middle of a piece of white card. Close your left eye and with your right eye fixate on a spot across the room and hold your gaze there. Now place the card in front of your face so that the dot lies in your line of sight and is in focus. Now move the card slowly to the right, maintaining your focus ahead. In about ten degrees of movement, the dot will disappear and then appear again after it has crossed the blind spot.

2 See Castner, 1983b,104-6; Castner and Eastman,1984; Chang et al.,1985; and more recently Steinke, 1987.

3 See Gibson (1950b, 368-9) and (1978, 229-33).

4 See, for example, his disclaimer (1978,228) in regard to his (1954), (1960), and (1966a, chap. 11) works. See also, Hamlyn's (1983) critique of Gibson's theory of perception.

CHAPTER 4

1 On Suzuki and his method, see Suzuki (1969), (1973), and Bigler and Lloyd-Watts (1979). Much of this section has been abstracted from these sources.

2 The full array of elements and activities in the Orff program can be seen in the guidelines for the three levels of training courses in Orff-Schulwerk. See AOSA*Guidelines*.

3   Analogies between sound and light have been made a number of times and in various ways since the time of Aristole (see Graves, 1952,6 and 89).
4   On Jaques-Dalcroze and eurthymics see Findlay (1971), Jaques-Dalcroze (1980), Wiggins (1986), and Vanderspar (n.d.).
5   R. Murray Schafer (1985), the Canadian composer, describes two mapping activities in downtown Basel, Suisse (Switzerland): a sound treasure hunt (*klangschnitzeljagd*) and a sound treasure walk with a commentary providing clues from the environmental soundscape for the discovery of a specific path through the streets.
6   See Cogan and Escot (1976) and their various graphs of music pieces in which graphic contrasts are made among, for example, the soprano, inner, and bass voices.
7   For other graphic techniques for describing aspects of dance see also Hawkins (1964), Kleinman (1968), Samuelson (1980), and Lewis (1984).

CHAPTER 5

1   David Melton (1985) also has invoked the "left brain-right brain argument" in encouraging teachers to have their young students illustrate their stories with interesting covers and binding them together in booklets.
In the geographic literature, Blake (1979) has noted the challenge of the "back to basics" movements to those disciplines that employ visual aids and makes a plea for "perceptual-conceptual maps," which seem to support the notion of less emphasis on formal mapping and more on the process of mapping.

CHAPTER 7

1   Board (1978) develops a similar list divided into "navigation, measurement, and visualisation" tasks.

Bibliography

ABBREVIATIONS

AAG	Association of American Geographers
ACSM	American Congress on Surveying and Mapping
AOSA	American Orff-Schulwerk Association
CIS	Canadian Institute of Surveying
FWTAO	Federation of Women Teachers' Associations of Ontario
ICA	International Cartographic Association
NAEA	National Art Education Association
NCGE	National Council for Geographic Education
NCSS	National Council for Social Studies
OACEE	Ontario Association for Geographic and Environmental Educators
Sixth Form	Sixth Form and Universities Working Group of the Geographical Association
SRI	Stanford Research Institute

AAG and NCGE. 1984. Guidelines for Geographic Education. AAG and NCGE.

Acredolo, Linda P. 1981. *Small- and large-scale spatial concepts in infancy and childhood.* In Liben, Patterson, and Newcombe (1981): 63-81.

Albarn, Keith and Smith, Jenny M. 1977. *Diagram: the instrument of thought.* London: Thames and Hudson.

Albers, Josef. 1971. *Interaction of color.* New Haven: Yale University Press.

Alexander, Hubert G. 1967. *Language and thinking: a philosophical introduction.*

New York: Van Nostrand.

AOSA. 1980. Guidelines for Orff-Schulwerk training courses, levels I, II, III. Cleveland: August Graphics for AOSA.

Argüelles, Jose. 1975. *The transformative vision*. New York: Random House.

Arnheim, Rudolf. 1974. *Art and visual perception*. Berkeley: University of California Press.

Balchin, W.G.V. 1972. Graphicacy. *Geography* 57, no. 3:185-95.

Balchin, W.G.V., and Coleman, Alice M. 1965. Graphicacy should be the fourth ace in the pack. *The Times Educational Supplement*, 5 Nov.

Bales, John, Graves, Norman, and Walford, Rex, eds. 1973. *Perspectives in geographical education*. Edinburgh: Oliver & Boyd.

Battersby, Martin. 1974. *Trompe l'oeil - the eye deceived*. London: Academy Editions.

Beard, Ruth. 1969. *An outline of Piaget's developmental psychology for students and teachers*. New York: New American Library.

Beller, Henry K. 1972. Problems in visual search. *International Yearbook of Cartography* 12:137-44.

Bettleheim, Bruno. 1980. Art and art education: a personal vision. In *Surviving, and other essays*, pp. 412-26. New York: Vintage Books.

Beyer, Barry K. 1985. Critical thinking: what is it? *Social Education* (Apr.):270-6.

Bigler, Carole, and Lloyd-Watts, Valery. 1979. *Studying Suzuki piano: more than music*. Athens, OH: Ability Development Associates.

Birkenshaw, Lois. 1986. Music for all children in the land of Orff and Kodaly. *Ostinato* (Carl Orff Canada) Bulletin no.32 (Jan.):12-17.

Blades, Mark, and Spencer, Christopher. 1986. Map use by young children. *Geography* 71, no.1:47-52.

Blake, William E. 1979. Right-left brain functions: geographic implications. *Journal of Geography* (Nov.):246.

Blaut, James M. 1987. Notes toward a theory of mapping behavior. *Children's Environmental Quarterly* 4, no. 4 (winter):27-34.

Blaut, James M., and Stea, David. 1969. *Place learning*. Place Perception Research Report no. 4. Worchester: Clark University, Graduate School of Geography.

– 1971. Studies of geographic learning. *Annals, AAG* 61:387-93.

– 1974. Mapping at the age of three. *Journal of Geography* 73(Oct.):5-9.

Blaut, James M., McCleary, George, and Blaut, America S. 1970. Environmental mapping in young children. *Environment and Behavior* 2:335-49.

Board, Christopher. 1978. Map reading tasks appropriate in experimental studies in cartographic communication. *Canadian Cartographer* 15, no. 1:1-12.

– ed. 1984. New insights in cartographic communication. *Cartographica* 21, no. 1: 138 pp.

Boardman, David. 1976. Graphicacy in the curriculum. *Educational Review* 28, no. 2 :118-25.

– 1983. *Graphicacy and geography teaching*. London: Croom Helm.

– ed. 1985a. *New directions in geographical education*. London: Falmer Press.

– 1985b. Spatial concept development and primary school map work. In Boardman (1985a): 119-34.

Brooks et al. 1983. Cognitive levels matching. *Educational Leadership* 40(May):4-8.

Broudy, H.S. 1983. A common curriculum in aesthetics and fine arts. In Fenstermacher and Goodlad (1983): 219-47.

Brown, John L. 1965. The structure of the visual system. In Graham et al. (1965): 39-59.

Bruner, Jerome S. 1962. *On knowing: essays for the left hand*. Cambridge: Harvard University Press.

– 1966. *Towards a theory of instruction*. Cambridge: Harvard University Press.

– 1978. Learning the mother tongue. *Human Nature* 1(Sept.):42-9.

Burke, James. 1978. *Connections*. Toronto: Little, Brown.

Butterworth, George. 1977. *The child's representation of the world*. New York: Plenum Press.

Carley, Isabel McN., ed. 1977a. *Orff re-echoes, book I*. Brasstown, NC: American Orff-Schulwerk Association.

– 1977b. On creativity. In Carley (1977a): 75-6.

– 1977c. That lovely two-headed Betsy Higginbottam. In Carley (1977a): 83-4.

– 1980. *My recorder primer: sing and play, book a*. Brasstown, NC: Brasstown Press.

– ed. 1985. *Orff re-echoes, book II*. Lakemont, GA: Copple House Books for The American Orff-Schulwerk Association.

Carpenter, Helen McC., ed. 1964. *33d yearbook NCSS*. Washington: National Council for Social Studies.

Carswell, Ronald J.B. 1971. Children's abilities in topographic map reading. In Castner and McGrath (1971): 40-5.

– 1986. Atlas skills for learning rather than learning atlas skills. *History and Social Science Teacher* 22, no. 1 (Fall):19-23.

Castner, Henry W. 1964. The role of pattern in the visual perception of graded dot area symbols in cartography. PH.D. dissertation, University of Wisconsin-Madison.

– 1973. Electro-oculography in cartographic research. Paper presented at the Annual Meeting, Canadian Association of Geographers, May, Thunder Bay, Ontario.

– 1979a. Viewing time and experience as factors in map design research. *Canadian Cartographer* 16, no. 2:145-58.

– 1979b. A model of cartographic communication: practical goal or mental attitude? *International Yearbook of Cartography* 19:34-40.

– 1980a. Might there be a Suzuki method in cartographic education? Paper presented at the 10th International Conference, ICA, Aug. Tokyo, Japan.

– 1980b . Special purpose mapping in 18th century Russia: a search for the beginnings of thematic mapping. *American Cartographer* 7, no. 2:163-75.

– 1981. Might there be a Suzuki method in cartographic education? *Cartographica* 18, no. 1:59-67.

– 1983a. Research in cartographic communication: any implications for introducing maps and mapping to young children? *The Operational Geographer* no. 2:13-16.

– 1983b. Research questions and cartographic design. In Taylor (1983), chapter 5: 87-113.

– 1984b. Twentieth century children's atlases: social force or educational farce? Invited paper presented at symposium, Images of the World: The Atlas Through History, sponsored by the Center for the Book and the Map Division of the Library of Congress, Oct.,Washington.

Castner, Henry. W., and Cromie, Brian W. 1982. Two perspectives on research on map design. CIS Centennial Convention, Ottawa. *Proceedings* 2(Apr.):122-41.

Castner, Henry W., and Eastman, J. Ronald. 1984. Eye-movement parameters and perceived map complexity - I. *American Cartographer* 11, no. 2:107-17.

– 1985. Eye-movement parameters and perceived map complexity -II. *American Cartographer* 12, no. 1:29-40.

Castner, Henry W., and Lywood, Denis W. 1978. Eye movement recording: some approaches to the study of map perception. *Canadian Cartographer* 15(Dec.):142-50.

Castner, Henry W., and McGrath, Gerald, eds. 1971. Map design and the map user. *Cartographica* monograph no. 2.

– 1984. Educating map publishers: evaluating changes in map design on the basis of map reading activities and visual tasks. *Technical Papers, 680-90.* 12th ICA, Perth, Australia.

Castner, Henry W., and Robinson, Arthur H. 1969. *Dot area symbols in cartography: the influence of pattern on their perception.* Technical Monograph #CA-4, 78pp. Washington: ACSM.

Castner et al. 1981. *Thinking about Ontario.* Toronto and Edmonton: Hosford Publishing.

Catling, Simon J. 1978. The child's spatial conception and geographic education. *Journal of Geography* (Jan.):24-8.

Chang et al. 1985. The effect of experience on reading topographic relief information: analyses of performance and eye movements. *Cartographic Journal* 22, no. 2:88-94.

Choksy, Lois. 1981.*The Kodaly context: creating an environment for musical learning.* Englewood Cliffs: Prentice-Hall.

Choksy et al. 1986. *Teaching music in the twentieth century.* Englewood Cliffs:

Prentice-Hall.

Cogan, Robert and Escot, Pozzi. 1976. *Sonic design: the nature of sound and music*. Englewood Cliffs, NJ: Prentice-Hall.

Collinson, Alan S. 1981. Is cartography in the doldrums? – a personal view. *Cartographic Journal* 18, no.1:58-9.

Common, Michael, Richards, Francis A., and Armon, Cheryl, eds. 1984. *Beyond formal operations: late adolescent and adult cognitive development*. New York: Praeger.

Common, Ron. 1985. From a speech to the Fall Meeting of OAGEE as quoted in the OAGEE Newsletter, Eastern Region, 19 Oct.

Cornsweet, Tom N. 1969. Information processing in human visual systems. *SRI Journal* Feature Issue no.5 (Jan.):16-27.

Coulter, Dee Joy. 1985. The brain's timetable for developing musical skills. In Carley (1985): 90-4.

– 1986. In defense of music. Audio Cassette. Longmont, CO: Coulter Publications.

D'Angelo, Edward. 1971. *The teaching of critical thinking*. Amsterdam: B.R. Gruner.

DeBono, Edward.1983. The direct teaching of thinking as a skill. *Phi Delta Kappan* 64(June):703-8.

DeLucia, Alan A. 1974. The map interpretation process, its observation and analysis through the technique of eye movement recording. PH. D. dissertation, University of Washington.

– 1976. How people read maps: some objective evidence. In *Proceedings*, 135-44. ACMS 36th Annual Meeting.

Dodwell, Peter C. 1960. Children's understanding of number and related concepts. *Canadian Journal of Psychology* 14:191-205.

– 1963. Children's understanding of spatial concepts. *Canadian Journal of Psychology* 17:141-61; also in Sigel and Hooper (1968).

Donaldson, Margaret. 1978. *Children's minds*. Glasgow: Fontana/Collins.

Dondis, Donis A. 1973. *A primer of visual literacy*. Cambridge: MIT Press.

Dowling, W. Jay, and Harwood, Dane L. 1986. *Music cognition*. Orlando: Academic Press.

Eastman, J. Ronald. 1985. Cognitive models and cartographic design research. *Cartographic Journal* 22, no.2:95-101.

Eastman, J. Ronald, and Castner, Henry W. 1983. The meaning of experience in task-specific map reading. In Taylor (1983), chapter 6: 115-47.

Edwards, Betty. 1979. *Drawing on the right side of the brain*. Los Angeles: J.P. Tarcher.

Ehrenzweig, Anton. 1969. *The hidden order of art: a study in the psychology of artistic imagination*. Berkeley: University of California Press.

Eisner, Elliot W. 1980. The role of the arts in the invention of man. *New York University Education Quarterly* (Spring): 2-7.

Ennis, Robert H. 1962. A concept of critical thinking. *Harvard Educational Review* 32(Winter):81-111.

Eric. *News Bulletin of the Clearinghouse for Social Studies/Social Science Education* (Winter 1985).

Eriksen, C.W. 1952. Locations of objects in a visual display as a function of the number of dimensions on which the objects differ. *Journal of Experimental Psychology* 44:126-32.

—1953. Object location on a complex perceptual field. *Journal of Experimental Psychology* 45:56-60.

Fenstermacher, Gary D., and Goodlad, John I. 1983. *Individual differences and the common curriculum.* 82d Yearbook of the National Society for the Study of Education. Chicago: University of Chicago Press.

Ferguson, Eugene S. 1977. The mind's eye: nonverbal thought in technology. *Science* 197, no.4306 (Aug.):827-36.

Fifty nursery songs within the range of five notes. New York: Boosey & Hawkes 1964.

Findlay, Elsa. 1971. *Rhythm and movement: applications of Dalcroze eurhythmics.* Evanston: Summy-Birchard.

Fish, Margaret. 1977. The value of the Orff approach. In Carley (1977a): 48-50.

Flavell, John H. 1963. *The developmental psychology of Jean Piaget.* Princeton: Van Nostrand.

Forrest, David, and Castner, Henry W. 1985. The design and perception of point symbols for tourist maps. *Cartographic Journal* 22(June):11-19.

Frankenstein, Alfred V. 1970. *The reality of appearances.* New York: New York Graphics Society.

Fraser, Dorothy M., and West, Edith. 1961. *Social studies in secondary schools.* New York: Ronald Press.

Freeman, James, Butcher, H.J., and Christie, T. 1971. *Creativity: a selective review of research.* 2d ed. London: Society for Research into Higher Education.

Freeman, Robert B. Jr. 1965. Ecological optics and visual slant. *Psychological Review* 72, no. 6:501-4.

Gaitskell, Charles D., and Hurwitz, Al. 1970. *Children and their art.* 2d ed. New York: Harcourt, Brace & World.

Gallwey, W. Timothy. 1974. *The inner game of tennis.* New York: Random House.

Garner, W.R. 1981. The analysis of unanalyzed perceptions. In Kubovy and Pomerantz (1981): 119-39.

Gazzaniga, Michael S. 1985. *The social brain: discovering the networks of the mind.* New York: Basic Books.

Gerber, Rodney. 1981. Young children's understanding of the elements of maps. *Teaching Geography* (Jan.):128-33.

Gibson, Eleanor J. 1969. *Principles of perceptual learning and development.* New York: Appleton-Century-Crofts.

Gibson, James J. 1950a. *The perception of the visual world.* New York: Houghton Mifflin.

– 1950b. The perception of visual surfaces. *American Journal of Psychology* 63:367-84.

– 1954. A theory of pictorial perception. *Audio-Visual Communication Review* 2, no.1:3-23.

– 1960. Pictures, perspective and perception. *Daedalus* 89, no.1:216-27.

– 1961. Ecological optics. *Vision Research* 1:253-62.

– 1965. Constancy and invariance in perception. In Kepes (1965): 60-70.

–1966a. *The senses considered as perceptual systems.* New York: Houghton Mifflin.

– 1966b. The problem of temporal order in stimulation and perception. *Journal of Psychology* 26(Mar):141-9.

– 1973. On the concept of 'formless invariants' in visual perception. *Leonardo* 6:43-5.

–1978. The ecological approach to the visual perception of pictures. *Leonardo* 11:227-35.

– 1979. *An ecological approach to visual perception.* Boston: Houghton Mifflin.

Gombrich, E.H. 1960. *Art and illusion.* London: Phaidon.

Goode, John P. 1922. *Goode's school atlas.* Chicago: Rand McNally.

Graham, Clarence H., et al., eds. 1965. *Vision and visual perception.* New York: John Wiley.

Graves, Maitland, E. 1952. *Color fundamentals.* New York: McGraw-Hill.

Guelke, Leonard, ed. 1981. Maps in modern geography: geographical perspectives on the new cartography. *Cartographica* 18, no.2: 213pp.

Hall, Doreen. 1960. *Music for children, teacher's manual.* Edition Schott 4898. New York: Schott Music Corp.

Hall, J. Tillman. 1963. *Dance! A complete guide to social, folk and square dancing.* Belmont, CA: Wadsworth.

Hamlyn, D.W. 1969.*The psychology of perception.* London: Routledge & Kegan Paul.

– 1983. *Perception, learning and the self.* London: Routledge & Kegan Paul.

Harris, Paul. 1977. The child's representation of space. In Butterworth (1977): 83-93.

Hart, Roger A. 1971. *Aerial geography: an experiment in elementary education.* Place Perception Research Report no.6. Worchester: Clark University, Graduate School of Geography.

Hawkins, Alma. 1964. *Creating through dance.* Englewood Cliffs: Prentice-Hall.

Hayes, J.R. 1976. It's the thought that counts: new approaches to educational theory. In Klahr (1976): 235-42.

Head, C. Grant. 1984. The map as natural language: a paradigm for understanding. In Board (1984): 1-32.

Herbert, Nick. 1987. *Quantum reality: beyond the new physics*. Garden City, NY: Anchor Press.

Holt, John C. 1964. *How children fail*. New York: Dell Publishing.

– 1969. *The underachieving school*. New York: Pitman.

– 1970. *What do I do Monday?* New York: Dutton.

Hooper, Frank H. 1968. Piagetian research and education. In Sigel and Hooper (1968): 423-34.

Howarth, Mary. 1984. Reading: the big hurdle for disadvantaged children. *FWTAO Newsletter* (Feb.):1-11.

Hughes, Martin. 1986. *Children and number: difficulties in learning mathematics*. New York: Basil Blackwell.

Jaques-Dalcroze, Emile. 1980. *Rhythm, music and education*. Translated by Harold F. Rubenstein. Rev. paperback ed. London: The Dalcroze Society.

Jarolimek, John. 1964. The psychology of skill development. In Carpenter (1964): 17-25.

Jenks, George F., and Caspall, Fred C. 1971. Error on choroplethic maps: definition, measurement, reduction. *Annals, AAG* 61(June):217-44.

Johns, Eunice, and Fraser, Dorothy McC. 1964. Social studies skills: a guide to analysis and grade placement. In Carpenter (1964): 310-27.

Jolliffe, R. 1974. An information theory approach to cartography. *Cartography* 8, no.4:175-81.

Julesz, Bela. 1975. Experiments in the visual perception of texture. *Scientific American* 232(Apr.):34-43.

–1981. Figure and ground perception in briefly presented isodipole textures. In Kubovy and Pomerantz (1981): 27-54.

Kahneman, D. 1973. *Attention and effort*. Englewood Cliffs: Prentice-Hall.

Kaufman, L., and Richards, N. 1969. Spontaneous fixation tendencies for visual forms. *Perception and Psychophysics* 5:85-8.

Keates, J.S. 1982. *Understanding maps*. London: Longman.

Kepes, Gyorgy. 1965. *The nature and art of motion*. New York: George Braziller.

Klar, David, ed. 1976. *Cognition and instruction*. Hillsdale, NJ: Lawrence Erlbaum.

Kleinman, Seymour. 1968. *Social dancing fundamentals*. Columbus, OH: Charles E. Merrill.

Kling, J.W., and Riggs, Lorrin A. 1971. *Woodworth and Schlosberg's Experimental psychology*. 3d ed. New York: Holt, Rinehart and Winston.

Kodaly, Zoltan. 1957. *24 Little canons on the black keys*. New York: Boosey & Hawkes.

Koestler, Arthur. 1964. *The act of creation*. London: Hutchinson.

–1968. Ethical issues involved in influencing the mind. Paper presented at the symposium on The University and the Ethics of Change. September, Queen's University, Kingston.

Koffka, Kurt. 1935. *Principles of Gestalt psychology*. New York: Harcourt Brace.

Kolacny, A. 1969. Cartographic information: a fundamental concept and term in modern cartography. *Cartographic Journal* 6:47-9.

Kracht, James B., and Boehm,Richard G. 1980. Geography is more than knowing: deciding and doing are basic too! *Journal of Geography* (Mar.):104-7.

Kubovy, Michael, and Pomerantz, James R. eds. 1981. *Perceptual organization*. Hillsdale, NJ: Lawrence Erlbaum.

Langer, Susanne K. 1957. Abstraction in science and abstraction in art. In *Problems of Art*, Ten philosophical lectures, 163-80. New York: Charles Scribner.

Leonard, George. 1977. *The ultimate athlete*. New York: Avon Books.

Lewis, Daniel. 1984. *The illustrated dance technique of José Limón*. New York: Harper & Row.

Liben, Lynn S. 1981. Spatial representation and behavior: multiple perspectives. In Liben, Patterson, and Newcombe (1981): 3-36.

Liben, Lynn S., Patterson, Arthur H., and Newcombe, Nora. 1981. *Spatial representation and behavior across the life span*. New York: Academic Press.

Liess, Andreas. 1966. *Carl Orff*. Translated by Adelheid and Herbert Parkin. London: Calder and Boyars.

Lindsay, John C. 1975. *Art is ...* . Agincourt, Ont.: General Learning Press.

Linstone, Harold A. 1984. *Multiple perspectives for decision making*. New York: North-Holland.

Litt, Lawrence. 1977. Naming of parts: how children describe and how children draw common objects. In Butterworth (1977): 73-80.

Lowenfeld, Viktor, and Brittain, W. Lambert. 1975. *Creative and mental growth*. 6th ed. 7th ed., 1982. New York: Macmillan.

Lynch, Kevin. 1960. *The image of the city*. Cambridge, MA: Technology Press.

McGrath, Gerald, ed. 1984. *The marketing of cartographic information*. Proceedings of the symposium, May, Queen's University, Kingston, Ontario.

McGregor, Brian R. 1988. The perception of reversed-line symbols on maps. MA thesis, Queen's University, Kingston, Ontario.

McKim, Robert H. 1972. *Experiences in visual thinking*. Monterey, CA: Brooks/Cole.

McPeck, John. 1981. *Critical thinking and education*. New York: St Martin's Press.

Mackworth, Norman H. 1965. Visual noise causes tunnel vision. *Psychonomic Science* 3:67-8.

Mackworth, Norman H. and Bruner J.S. 1970. How adults and children search and recognize pictures. *Human Development* 13:149-77.

Marr, David. 1982. *Vision: a computational investigation into the human repre-*

sentation and processing of visual information. San Francisco: W.H. Freeman.

Matthews, Hugh. 1986. Children as map-makers. *Geographical Magazine* 58, no. 3:124-6.

Melton, David. 1985. *Written and illustrated by* ... . Kansas City: Landmark Editions.

Meyer, Judith M.W. 1973. Map skills instruction and the child's developing abilities. *Journal of Geography* 72(Sept.):27-35.

Ministry of Education. 1975. *The formative years*. Circular P1J1, Provincial Curriculum Policy for the Primary and Junior Divisions of the Public and Separate Schools of Ontario. Ontario: Ministry of Education.

Mitroff, Ian I. 1984. In foreword to Linstone (1984): xiii-xvii.

Morrison, Joel L. 1976. The relevance of some psychophysical cartographic research to simple map reading tasks. Paper presented at the 8th International Conference, ICA, Moskva.

Muehrcke, Phillip C. 1976. Concepts of scaling from the map reader's point of view. *American Cartographer* 3:123-41.

– 1981. Maps in geography. In Guelke (1981): 1-41.

Muller, Jean-Claude. 1976. Objective and subjective comparison in chorople-thic mapping. *Cartographic Journal* 13, no. 2:156-66.

Munby, A. Hugh. 1982. *What is scientific thinking?* A Discussion Paper, Science Council of Canada. Ottawa: Minister of Supply and Services.

Munby, A. Hugh, and Russell, Thomas. 1983. A common curriculum for the natural sciences. In Fenstermacher and Goodlad (1983): 160-85.

NAEA. n.d. A position statement by NAEA. Washington, DC:NAEA

Neisser, Ulric. 1967. *Cognitive psychology*. Englewood Cliffs, NJ: Prentice-Hall.

– 1976. *Cognition and reality: principles and implications of cognitive psychology*. San Francisco: W.H. Freeman.

Nelson Atlas of Canada. 1978. Toronto: Nelson Canada.

Nelson, William L. 1980. An analysis of the effects of texture on the periph-eral perception of map symbology. MA thesis, Queen's University, King-ston, Ontario.

O'Brien, James P. 1989. Integrating world music in the music"appreciation" course. *Ostinato* bulletin no. 42 (Apr.):8-9.

Papert, Seymour. 1980. *Mind-storms*. New York: Basic Books.

Parrott, Dorothy. 1986. Developmental perceptions. Oral presentation at the conference Music, Movement, Brain, Englewood, CO.

Peniakoff, Vladimir. 1950. *Private army*. London: Jonathan Cape.

Petchenik, Barbara B. 1977a. Mapping the eighteenth century. *Scholarly Publishing* 8(July):357-66.

– 1977b. Cartography and the making of an historical atlas. *American Cartog-rapher* 4(Apr.):11-28.

– 1979. From place to space: the psychological achievement of thematic mapping. *American Cartographer* 6(Apr.):5-12.

– 1983. A map maker's perspective on map design research 1950-1980. In Taylor (1983): 37-68.

– 1984. Facts or values: basic methodological issues in research for educational mapping. In *Technical Papers* 789-804. 12th International Conference, ICA, Perth, Australia.

Piaget, Jean. 1950. *The psychology of intelligence*. London: Routledge & Kegan Paul.

– 1954. *Construction of reality in the child*. New York: Basic Books.

Piaget, Jean, and Inhelder, Bärbel. 1956. *The child's conception of space*. New-York: Humanities Press.

Pick, Anne D. ed. 1979. *Perception and its development: a tribute to Eleanor J. Gibson*. Hillsdale, NJ: L. Erlbaum Associates.

Pirsig, Robert M. 1981. *Zen and the art of motorcycle maintenance*. New York: Bantam Books.

Platt, J.R. 1968. The two faces of perception. *Main Currents* 25, no.1:8-19.

Polanyi, Michael. 1959. *The study of man*. Chicago: University of Chicago Press.

Proctor, Nigel. 1984. Geography and the common curriculum. *Geography* 69, no.1:38-45.

Randhawa, Bikkar S. 1987. Atlases for children: a legacy of perceptual and cognitive processes. In Atlases for schools: design principles and curriculum perspectives, ed. R.J.B. Carswell, G.J.A. de Leeuw, and N.M. Waters. *Cartographica* 24, no.1:47-60.

Ratajski, Lech. 1973. The research structure of theoretical cartography. *International Yearbook of Cartography* 12:217-28.

Reed, Edward S. 1988. James J. Gibson and the psychology of perception. New Haven: Yale University Press.

Resnik, Lauren B. 1976. Task analysis in instructional design: some cases from mathematics. In Klahr (1976): 51-80.

Riggs, Lorrin A. 1971. Vision. In Kling and Riggs (1971): 273-314.

Robinson, Arthur H. 1971. Discussion: an interchange of ideas and reactions. In Castner and McGrath (1971): 46-53.

– 1985. A classification of maps for the history of cartography. Paper presented at the 11th International Conference on the History of Cartography, 8-12 July, Ottawa, Ontario.

Robinson, Arthur H., and. Petchenik, Barbara P. 1975. The map as a communication system. *Cartographic Journal* 12:7-15.

– 1976. *The nature of maps*. Chicago: University of Chicago Press.

Robinson, Arthur H., Sale, Randall, and Morrison, Joel. 1978. *Elements of cartography*. 4th ed. New York: John Wiley.

Rosenfeld, Anne H. 1985. Music, the beautiful disturber. *Psychology Today* (Dec.):48-56.

Rushdoony, H.A. 1968. A child's ability to read maps: summary of the research. *Journal of Geography* 67:213-22.

Russell, David H. 1956. *Children's thinking*. Boston: Ginn & Co.

Salome, Richard A. 1965-6. Perceptual teaching and children's drawings. *Studies in Art Education* 7, no. 1:18-33.

Salome, Richard A., and Reeves, D. 1972. Two pilot investigations of perceptual training of four- and five-year old kindergarten children. *Studies in Art Education* 13, no. 2:3-9.

Samples, Bob. 1976. *The metaphoric mind: a celebration of creative consciousness*. Reading, Mass.: Addison-Wesley.

Samuels, Mike, and Samuels, Nancy. 1975. *Seeing with the mind's eye*. New York: Random House.

Samuelson, Miriam. 1980. About dance notation. In *Music for children*. American ed. Vol. 3:333-4. Valley Forge, PA: Schott Music Corp.

Satterly, David J. 1973. Skills and concepts involved in map drawing and map interpolation. In Bales, Graves, and Walford (1973): 162-9.

Sauvy, Jean, and Sauvy, Simonne. 1974. *The child's discovery of space*. Baltimore: Penguin Books.

Schafer, R. Murray. 1985. Experience in Basel. In Carley (1985): 239-43.

Schiffman, H.R. 1976. *Sensation and perception: an integrated approach*. New York: John Wiley.

Schrage, Michael. 1985. Artificial intelligence: teaching computers power of creative stupidity. *Washington Post*. Sunday, 1 Dec.

Science 36. *Science for every student: educating Canadians for tomorrow's world*. Summary of the larger Report no. 36 of the same title. Ottawa: Science Council of Canada. 11pp.

Science 52. 1984. *Science education in Canadian schools*, by Graham W.F. Orpwood and Jean-Pascal Souque. Summary of the three volume Background Study 52 of the same title. Ottawa: Science Council of Canada. 26pp.

Seiderman, Arthur, and Schneider, Steven. 1983. *The athletic eye*. New York: Hearst Books.

Shamrock, Mary. 1986. Orff Schulwerk: an integrated foundation. *Music Educator's Journal* (Feb.):51-5.

Sigel, Irving E., and Hooper, Frank H. 1968. *Logical thinking in children*. New York: Holt, Rinehart and Winston.

Singer, Dorothy G., and Revenson, Tracey A. 1978. *A Piaget primer: how a child thinks*. New York: International Universities Press.

Sixth Form. 1979. Skills and techniques for sixth-form geography. *Geography* 64:37-45.

Smith, Ralph A. 1978. Critical reflections on the AGE idea. *Music Educator's Journal* 64, no.5:88-97.

Sorrell, Patrick. 1974. Map design – with the young in mind. *Cartographic Journal* 11(Dec.):82-90.

Southworth, Michael, and Southworth, Susan. 1982. *Maps: A visual survey and design guide*. Boston: Little, Brown and Co.

Steinke, Theodore R. 1987. Eye movement studies in cartography and related fields. In Studies in Cartography: A Festschrift in Honor of George F. Jenks, ed. Patricia Gilmartin. *Cartographica* 24, no.2:40-73.

Stroebel, Leslie, Todd, Hollis, and Zakia, Richard. 1980. *Visual concepts for photographers*. New York: Focal Press.

Suzuki, Shinichi. 1969. *Nurtured by love: a new approach to education*. Translated by Waltraud Suzuki. New York: Exposition Press.

–1973. *The Suzuki concept: an introduction to a successful method for early music*. Berkeley: Diablo Press.

Taylor, D.R. Fraser, ed. 1983. *Graphic communication and design in contemporary cartography*. Toronto: John Wiley.

Thorndike, R.L. 1950. How children learn the principles and techniques of problem-solving. In *49th Yearbook*, part 1, 192-216. National Society for the Study of Education.

Time. 9 May 1983. To stem a 'tide of mediocrity,' 61-3.

Vanderspar, Elizabeth. n.d. Dalcroze handbook: principles and guidelines for teaching eurhythmics. Mimeo.

Wadsworth, Barry J. 1971. *Piaget's theory of cognitive development*. New York: David McKay.

Watkins, C. Law. 1946. *The language of design*. Washington, DC: Phillips Memorial Gallery.

Weikart, Phyllis S. 1981. *Movement to the Musica Poetica*. St Louis: Magnamusic-Baton.

Wiggins, Patti. 1986. Interview with Robert M. Abramson. *The Orff Echo* 18, no.3(Spring):2-3,19.

Williams, Leon G. 1967. The effects of target specification on objects fixated during visual search. *Acta Psychologica* 27:355-60.

–1971. Obtaining information from displays with discrete elements. In Castner and McGrath, eds. (1971): 29-34.

Wilson, Marjorie, and Wilson, Brent. 1982. *Teaching children to draw: a guide for teachers and parents*. Englewood Cliffs: Prentice-Hall.

Wolter, John A. 1975. Cartography – an emerging discipline. *Canadian Cartographer* 12:210-16.

Wuytack, Jos. 1977. Didactic principles in music education. In *Music for Children: Carl Orff Canada*, translated by Doreen Hall, bulletin no.7 (Sept.):3-7.

Yarbus, A.L. 1967.*Eye movements and vision*. New York: Plenum Press.

Index